STEAM TOYS

A Symphony in Motion

MORTON A. HIRSCHBERG

Schiffer Publishing Ltd
77 Lower Valley Road, Atglen, PA 19310

Dedication

This book is dedicated to my wife, Françoise, who told me "You should buy one good toy a year"–and then let me greatly exceed that number.

Published by Schiffer Publishing, Ltd.
77 Lower Valley Road
Atglen, PA 19310
Please write for a free catalog.
This book may be purchased from the publisher.
Please include $2.95 postage.
Try your bookstore first.

We are interested in hearing from authors
with book ideas on related subjects.

Printed in Hong Kong

ISBN: 0-7643-0009-1

Library of Congress Cataloging-in-Publication Data

Hirschberg, Morton A.
Steam toys: a symphony in motion/by Morton A. Hirschberg.
p. cm.
ISBN 0-7643-0009-1
1. Mechanical toys--Collectors and collecting.
2. Steam-engines--Equipment and supplies--Collectors and collecting. I. Title.
TS2301.T7H56 1996
688.7'28--dc20 96-18028
CIP

Table of Contents

Acknowledgements

I extend my gratitude to all those who inspired me, helped assemble my collection, or otherwise provided assistance in this project. First, to Dr. Ed Hoffman. If I had not seen the toys on your wall I would not be so crazy now; some of your toys are now on my wall. Second, to Louis H. Hertz, whose prose in *"The Handbook of Old American Toys"* made me realize that steam toys really are, as he wrote, a "symphony in motion." Third, to the four people who significantly influenced my collection: Frank Whitson, Doug Heinmuller, Rolf Naskret and Klaus Grutzka. Fourth, to Bill and Jeanne Bertoia, Carl Lobel, Louis Natenshon and Brent Rowell for providing photographs of toys one dreams of. Fifth, to Dr. Rurik Loder for translating some difficult German passages. Finally, to the many friends, collectors, and dealers who helped by adding one or more pieces: Tim Turner, Ed and Mickey Hirsch, Bob and Marianne Schneider, Jay and Becky Lowe, Jim Yeager, Stewart Waldman, Gary Cenname, Gus and June Majeune, Larry Bruch, Alan Green, Bill and Stevie Weart, Bill Bertoia, Noel Barrett, Joe Freeman, Carl Lobel, Tom Sage, Athel and Kathy Spilhaus, Russ Harrington, Curtis Smith, Sam Downey Sr., Sam Downey Jr., Steve Leonard, Ed O'Leary, Robert Rioux and Klaus Banke.

To Peter Schiffer for approving the project. To Mark S. Balbach and Richard Breneman, who, unless otherwise noted, photographed the collection. Most important, to Leslie Bockol, my editor, for her perceptive comments and suggestions which greatly improved the manuscript.

Thank you all.

Introduction

My grandmother stepped on, kicked, gave away, or threw out all my toy trains. My mother introduced me to mechanical banks via the nicely reproduced *Book of Knowledge* series. Gaining a conditioned dislike to the former and an indifference to the latter (including original banks) opened a wide new door into the total realm of toys and toy collecting. My love of cast iron novelty and bell toys gradually transformed itself into a deep passion for the charming and fanciful toys powered largely by steam. My immediate interest was in those articulated, colorful tin toys with pulleys on them. No engines, no boats, no trains, no tractors, no planes, no cars, and certainly no wind-ups, please–just those toy models known as accessories. And my interest, focus and resolve has persevered.

A True Story

Neither I nor my wife considered ourselves specialty collectors, although I once bought a number of steins and (when we tried our hand as antiques dealers) old tools. Early in 1980 after breakfast one morning, I said to my wife, "Let's buy old toys."

"What do you mean?" she asked.

I replied, "Not trains, not mechanical banks, and certainly not fire engines."

"Well then, what?" she asked.

I answered, "I don't know, but we will know them when we see them."

"And where will we find them?" she asked.

I reminded her of a shop in Baltimore, run by Frank Whitson, which we had seen after moving to Maryland, and off we went. And see them we did.

When we saw Ed Hoffman's collection two years later, in 1982, I began to warm up to tin toys. There were many, hanging on the wall seemingly suspended in air–so colorful, so articulated. There were more on a circular iron staircase. They were there, ready for motion–ready for the symphony.

I bought my first steam toy (mind you, not an accessory but a toy) at auction early in May 1984. A few weeks later we went to Europe and I found ten more at the Paris and London toy shows. I was able to pick up a copy of the Battenberg book on steam toys by Baecker and Kaiser, and dreamt of owning each of the toys pictured. With great zeal, passion, and aggressiveness I have been able to acquire many of them, and own the majority of the toys pictured here.

In 1992, Peter Schiffer asked if there was enough interest in steam toys to warrant a book. I told him I thought there *were* a lot of interested collectors but that more importantly, these toys had to be seen in their proper light, their true light–as toys, not just steam accessories. I also told him that someday I would write a book, but that I was not yet ready. Early in 1994 Peter called again, and said he was ready if I was. This book is the result of that conversation.

The ideal book would be an encylopedic work with manufacturers and trade catalogs (usually line drawings) juxtaposed with pictures of the actual toys. One would then be able to see and contrast the conceptualizations with the production pieces. Again, ideally, complete documentation of each toy via patents as well as manufacturing procedures and a history of each manufacturer would be presented.

In reality there is much information that is not known about many of the toys that were manufactured. Original catalogs are not readily available. Reprinted catalogs, which are a rich source of information about many toys, are themselves often incomplete. Furthermore, information is not available for all of the manufacturers.

In the pages that follow, I have tried to present a comprehensive sampling of steam toys. Toys are listed alphabetically by manufacturer, with a chapter devoted to each manufacturer. I give the name of each toy (often followed by the manufacturer's German name for it), manufacturer's identification (if known), earliest known date of production, a description or short narrative (if warranted), secondary information (e.g. dual manufacturers or other pertinent data), measurements (in centimeters), a range of prices, and finally, a provenance (if any). Some toys had a long production run, but in general the date of manufacture does not affect value if the toy remained unchanged over the run. Each toy has been carefully measured. Dimensions are given as length, width, and height as seen from the front of the toy. The price range covers what one might expect to pay a dealer for a piece in the condition shown. I have endeavored to provide as much information as possible, as correctly as possible; neither deceit nor harm is intended if the identification is or proves to be wrong.

A short discussion of each manufacturer whose toys are illustrated is provided. In some instances, more comprehensive material about a manufacturer may be found in the references listed. I also provide some information about the manufacturer's mark or logo. Several manufacturers changed their marks over time. Knowing when such changes occurred can assist in dating a toy, but be aware that many toys bear no marking or labels.

I have decided not to include information about manufacturers if their toys are not pictured. To the best of my knowledge, this is not a serious oversight. For example, a Lutz or Rock & Graner steam toy is fairly rare and is not discussed. Also omitted are Emil Hausmann and Bernard Hommola, as well as those German, English and American firms who produced steam engines primarily. For similar reasons, I have provided no information on major retailers, whose effect on the promotion and sales of steam toys and accessories can not be underestimated: Bassett-Lowke Ltd.; Gamages Ltd.; Marshall Field; F.A.O. Schwarz; Sears, Roebuck & Company; and Ullmann and Englemann.

A list of many references is provided. The Battenberg reference provides the most information on steam toys and is far and away the best reference on steam engines. Each, however, has proved useful to me and contains information on steam toys. A note of caution: misinformation seems to follow one of Newton's Laws. Once in print, it persists.

We know new information is found more often than one would imagine. In the last decade, several new catalogs have been found and new information provided. Readers are encouraged to provide information for future editions of this book. Information may be sent to the publisher.

Finally, perhaps unlike other references, this book shows examples of many of the more plentiful toys in addition to the rarer pieces. We are aware of many examples which are not illustrated, but by no means know of every single example ever manufactured. Indeed, for every toy on the wish list, about ten previously unknown or unseen toys appear.

May you find your toys and your toy dreams pictured here. May the symphony begin for you now.

Prices

"Ain't no damn pulley toy worth more than a hundred bucks."
-Frank Whitson

Price guides: everyone but publishers (no exception here) hates them.

Still, it is important to know that toy prices have risen sharply over the past few decades. There have been perturbations that followed the stock market, but to a lesser extent and with a quicker recoveries. Fine arts dealers extol the virtue of rare toys as pieces of folk art. Speculators flit in and out of the toy market. Toy prices reach record heights with each auction. Toys are in. Toys will stay in. Toy prices will remain strong.

It is my personal feeling that among all the categories of toys, steam toys (unless extremely rare) are grossly undervalued.

A True Story

A short time ago a friend of mine and fellow toy collector and I had the following conversation. He, "If you ran across a rare toy and it was only in fair condition but reasonably priced, would you buy it?"

Me, "Yes."

He, "Me too!"

A True Story

At a toy show in Macungie, Pennsylvania several years ago I spotted a fairly clean Carette toy with two men hammering at an anvil. I negotiated the price, which at the time was substantial. After I purchased it, the dealer asked me if I knew who had made it. When I told him "Carette," his face dropped. I assured him that even though it was a Carette (and he had doubts) he was getting a hefty sum. For me, the merits of the toy rather than the manufacturer per se are factors in determining prices.

The prices listed here are based on general and specific knowledge. They reflect rarity and condition. They are also based on my observations of prices obtained at auction and quoted by dealers. In many instances where examples are scarce to rare, toys may be sold privately, or, if at auction, true market prices cannot be firmly established.

There are some color variations, but unlike several other categories of toys, prices are not affected by color variation.

A True Story

At an Atlantic City show shortly before his death, Frank Whitson told me there was a small Mickey Mouse (slate) dancer for sale. He said he would have bought it for me, but at $1700 he thought it was too much. I told him I thought he was mistaken and that the asking price was more likely $17,000.

"Can't be," he said.

I told him I would check and let him know. I did so and, indeed, the asking price was $17,000. When I told Frank what I found out he said "Well ain't that somethin. Ain't no damn pulley toy worth more than a hundred bucks."

My advice is to buy what you like. Buy the best toys you can afford. Buy the toys in best condition. Buy for enjoyment. Remember that toy collecting, unlike many other pastimes, leaves you with something of value even if you pay a little too much. In addition, the market may pleasantly surprise you.

Finally, the prices listed here are in no way meant to set a standard or be used as an absolute measure. The author assumes no liabilities and means no harm.

Identification

"It's a Märklin"

When a seller is unsure of the maker, he is most apt to say "It's a Märklin." However, the surest identification arises when a toy bears a mark, stamp, or label. Gwen White's book, *Toys, Dolls, Automata: Marks and Labels,* is a comprehensive source for identifying toy manufacturers. In later chapters I include either pictures or drawings of marks which may possibly be found on toys. About half of all steam toys are not marked. Very few carousels or Ferris wheels are marked.

Manufacturers' catalogs and reprints are the next best source for mark and label information, even if they provide only line drawings of the toys. Still, be careful: as we now know, suppliers to several of the major manufacturers produced near look-alikes, often for sale through their own firms. In addition, several manufacturers seem to have copied designs closely from each other.

Measuring Steam Toys

There are clues other than markings which can help to determine the true manufacturer. One of these is the size of the toy. I have found that line drawings from three or four manufacturers are identical, but the actual toys are of different sizes.

A note on measurement: all the toys pictured in this book were measured as accurately as possible. Measurements are in centimeters. When there is a circular base, the diameter is given, denoted by a "d" following the numeric value. Additional measurement may be given if the top of the toy is larger than the base. In general, measured heights do not include the height to a flag if one is pictured. Otherwise, dimensions do reflect the greatest dimension observed (which might include a pulley).

I first measured my pieces to confirm dimensions listed in manufacturers catalogs. Much to my surprise, the listed dimensions were slightly greater than what I measured. That is, if I measured 11.6 the catalog would usually show 12. I speculate that manufacturers were paying their printing costs by the character, and thus could save money by rounding. And why not–who was going to measure and complain about a toy they paid a few pennies for! Bing catalogs do show some smaller dimensions (the true dimensions) in later catalogs, but this is not universal. I do not believe this means their toys were larger early on; maybe they had a few complaints about exaggerated measurements after all!

Comparing Toys

Barring identification through the means described above, one must look for points of similarity between the toy to be identified and a known toy. Of course, the more points of similarity, the better. Things to look for are similar figures, sizes, shapes, and color. Manufacturers were fairly specific in the shapes of the figures used in their toys. One problem does arise here, however: corporate takeover. Corporate takeover is not a new phenomena. Though one company might be bought out by another, the production processes did not change. Still, this might be the best available method of identification.

Another way to identify toys is to look for similarity in the bases of the toys. Manufacturers had distinct bases and supports. Note the number of levels of beveling on a base, and types of perforations (whether diamond or round, or top or bottom). Next, note the construction of roofs, girders and riveting. Observe the construction of pulleys, including the number and shape of the spokes. Look at the thickness of the tin.

Finally, look for color and general ornamentation. Manufacturers often painted using certain colors and color schemes. Note, for instance, how the color on the girders or supports of Doll Ferris wheels changes from top to bottom. This is characteristic of Doll.

Some Confusing Issues

There are a few other points to consider when you are identifying steam toys. First, new information has come to light regarding who manufactured what. Apparently, the Zschopau manufacturers produced a great number of pieces for Bing, Carette and Plank. In addition, most of the steam toys once attributed to Karl Bub are now attributed to the Zschopau School.

Second, as a company went out of business, one or more other firms would often pick up portions of their production line (if not the entire line). Besides the Zschopau School just mentioned, many other manufacturers share similarities. This is evident between Arnold and Bing at the turn of the century, between Arnold and Falk around 1930, and between Bing and Falk in the 1930s. Indeed, Arnold did supply toys to Falk.

Finally, there is the phenomena of persistence of misinformation (of which I hope I am not overly guilty myself). Toys misidentified in early references continue to be misidentified in newer writings.

A True Story

While I have been largely successful in properly identifying the toys pictured here, I did have to change some of my initial identifications. This was true of the toys of the Zschopau School. I asked my wife, Françoise, to see which toys she thought were similar. In an instant she noticed that the Becker workmen had their right legs forward and the Wunderlichs their left. Well done madam, well done. However, there are at least six different body styles between the two makers (barrel chests excluded). Leg position is a valuable aid but not definitive.

Confusion between Bing and Falk Ferris wheels, mills, and windmills can cause misidentification as well. I have attempted to properly identify the examples shown here. Color and size variations, though subtle, provide clues. Similarly, many Arnold toys are attributed to Bing. Be careful here, since Bing marked many of their toys and Arnold did not. Close observation shows that most of the full-figured Arnold workmen wear shop aprons, as do a handful of the early Bing figures; however, the Bing figures are thinner their aprons are often painted gray. It may be speculated that Arnold was supplying some toys to Bing. In another instance, clues to identification of the full-figured early Carette workmen is their waist-length, Eton-style jackets. In any event, the prices of many plentiful toys should not be based solely on the presumed manufacturer.

I have found that when a manufacturer is not known, the order of attribution is: Märklin, Bing/Carette, and Plank. *BEWARE!*

What's out there?

> ## A True Story
>
> *At a toy show a dealer asked me what I collected. I replied, "Steam accessories."*
>
> *"You mean like all those little men sawing?" the dealer asked.*
>
> *"Yes." I replied. "Exactly like all those little men sawing."*

A fair question to answer is 'What's out there?'

At first blush it seems if there are only "little men sawing," windmills, and fountains. In fact, there is a great variety and even strong crossover appeal among collectors. A sampling:

Workmen: blacksmiths, woodcutters/sawyers, tailors, cobblers, butchers, bootblacks, grinders, coopers, pumpers, firemen and scissors sharpeners.

Mills: watermills and windmills, with and without workmen and animal figures. These prove very popular with train collectors for their layouts.

Fountains: These are also popular with train collectors.

Musicians: Organ grinders, violinists, harpists.

Sports: Tennis players, boxers, wrestlers.

Carousels and Ferris Wheels: These run a full gamut and have great crossover appeal with carnival/circus toy collectors as well as general toy collectors.

Animals: Dogs, cats, elephants. These have great crossover appeal with general toy collectors.

Clowns: Great crossover appeal with carnival/circus collectors.

A quick count of toys in just four catalogs (Becker, Carette, Märklin, and Doll) showed over six hundred steam toys. I estimate there are about two thousand different steam toys, including transmissions but not counting engines, boats, trains, tractors, planes, or cars.

It should be noted that many toys made as wind-ups were also manufactured as steam toys.

Toy Names

Just like other types of toys, many steam toys were given names by their manufacturers. Usually, these were self-evident: "Sawyer," "Blacksmith," etc. In a few instances more colorful names were given. For instance, the Carette figure whose shoes are being shined is called (in German) *"Gigerl."* There is no good one-to-one translation for this word. At best, it means "a dandy" or "a fop." At worst, it is something akin to the evil gang members in *A Clockwork Orange* (notice the walking stick!). Could a *gigerl* be a fool with a mean streak?

I have tried to find out why the Ferris wheels and merry-go-rounds are called "Russian Carousels" and "Russian Swings." Ferris was an American inventor and his "wheel" was submitted as a design for the Chicago World's Fair of 1892-1893, the Columbian Exposition, to rival the Eiffel tower. Though not initially liked by the Fair's sponsors, the wheel was an immediate sensation with the public. My best conjecture is that "Russian Carousels and Swings" were so named because of the actual size of the carnival rides. That is, their large size was equated to the size of the land mass of Russia! The name of the carnival ride was carried over to the toys.

Cleaning Toys

Extreme care must be taken in cleaning any toy. I recommend that the reader research as many methods as possible before trying any of them. I bear no responsibility for any methods cited. May the cleaner beware.

The major problems involve the paint. First, pressure will remove any paint that already is or looks crazed. Many toys were not primed and flaking is common. Second, even if primed, many toys were painted with soluble paint.

In spite of this, I do try to clean my toys, using Jubilee furniture polish. First, I test a small portion very carefully using a cotton ball. I try to find an area on the back of the toy or underneath to test the effect. The Jubilee will also leave a bit of a protective coat as well as a shine.

I have tried the very expensive #7 car polish, but have not had results as good as with the furniture polish.

There is a commercial product sold at train shows but I have never used it. In addition, some folks like to spray on WD40 and gently wipe off excessive moisture. I have never tried that.

DO NOT USE ANY ABRASIVE CLEANERS.

Once I have cleaned them, I leave my toys alone. I don't even try to dust them.

Condition

We all want toys in the best condition that can be found: mint, if possible. For toys which had a long run, are easier to find, or are of lesser value, I suggest buying them as near to mint condition as possible. As the rarity increases, one must decide what one can tolerate. I myself have not repainted any toys but have bought toys that are completely repainted. I have, in addition, had minor repairs made by a professional restorer. Finally, I have added new flags and composition figures when necessary to complete a toy and bring it close to its original look. I have absolutely no qualms about doing this and also feel that flags and figures greatly enhance the feel and charm of a toy.

A True Story

Shortly after I purchased one of my smaller treasures from Frank Whitson at a major toy show, he said "I'm sure glad you bought that toy. I knew you would want it." Then, as a loud aside, clearly audible by all within near shouting distance he said, "I thought I would never get rid of that piece of junk."

Buying

Toys can be purchased from dealers and collectors privately, at toy shows, and at auction. When it is impossible to inspect a toy personally, it is helpful to request pictures. All modes of purchase seem equally satisfactory. Although I can delay gratification when waiting for a toy to arrive through the mail, nothing beats the rush of buying at auction; the hammer never seems to fall quickly enough for me.

A True Story

Some years ago at a local meet of the Train Collectors of America, I was witness to quite a collecting ploy. A young dealer ("YD") was holding a toy for a prominent collector ("PC"). A well-known dealer ("WKD") walked by and spotted it on his table. "How much do you want for that toy?" asked WKD.

"Oh, I'm holding it for PC," replied YD.

"Haven't you heard?" asked WKD loudly. "PC lost all of his money on the stock market. Doesn't have a dime to his name. In fact," he added, "my wife and I take lunch packages to his house almost every day."

The surprised YD quickly sold the toy to WKD, who then left.

My surprise over this quickly turned to amusement since PC was far from being broke and destitute. I decided to wait and see what would happen when PC showed up to claim his reserved toy. I did not have long to wait; PC came strolling by. "Where's my toy?" he asked.

YD answered, "I sold it to WKD"–and quickly explained the situation. I nodded in agreement. PC showed admirable restraint and set off to find WKD.

By this time, WKD already had the toy set up on his table, at double the price he had just paid. As PC approached, WKD remarked, "I thought you might like this toy."

"You know", said PC, "YD was holding that toy for me."

"Well, give me what I am asking and it's yours," *said WKD.*

Try as he may, PC could not strike a better deal. Eventually, he left.

That toy stayed on WKD's shelf for many months, until finally, I bought it–at the inflated price, of course.

The true story about the prominent collector ("PC") and his rival well-known dealer ("WKD") is not as mean-spirited as it may sound. WKD and PC were actually close friends. The final sale price was, in reality, quite reasonable. This episode was gamesmanship at its finest.

I don't think there is a moral to this story, but it does present a dilemma. Prominent collectors want dealers to hold merchandise for them. No one wants to hear dealers say that they just sold the best piece of merchandise they've ever handled just minutes ago. However, dealers want to move their merchandise, especially when it is easily salable. They own their merchandise and can sell it to whomever they please.

Education

A large portion of collecting has been the process of learning about collectibles. The public library is an inadequate source for the advanced collector, so I, like many other collectors, have acquired an extensive personal library. In addition to books there are many catalogs and reprints which are available. Today there are also many periodicals devoted exclusively to toys. *The Antique Toy World* published by Dale Kelley is my personal favorite, but much information on old toys also appears in *Inside Collector* and *Collector's Showcase.*

A True Story: Mad Dogs and Englishmen

A few years ago I was in Portsmouth, England at a music recital and struck up a conversation with the woman sitting next to me. When she asked if I spent my time listening to music, I had to tell her that I was usually out looking for a certain type of toy. When she pressed me for details, I told her that I collected steam toys–and she asked "Does Märklin mean anything to you?"

I replied enthusiastically, and inquired how she knew about Märklin. She told me that her husband, like many other Englishmen, had a number of engines and accessories in his shed. She went on to say no self-respecting Englishman would be without an engine and a toy or two in his shed.

Englishmen mad?

I don't think they're mad at all!

Some Speculations

Some writers and theorists speculate that toys served as models and prototypes for larger inventions. I do not fully subscribe to such speculation, since the reverse is also true, but we know that patents could often not be obtained without a working model. In general, collectors are interested solely in manufactured pieces and not models or prototypes, especially if the latter are one-of-a-kind.

Looking at steam toys reveals only a few themes: hard-working men and women at their daily tasks, machine tools, fountains, mills, and lastly, fanciful carnival toys like Ferris wheels and carousels for pleasure. In general, life in the days of the steam toy was hard, and the toys reflect that.

If one looks at the industrialization of a country, one can almost predict their progress and might in the world order by examining the toys they manufactured. Historically, toy manufacturing is about a generation or twenty years before emergence as a superpower. France and England were already superpowers but began extensive production in the 1840s and 1860s, respectively. The United States started in the 1870s, Germany just before the turn of the century, and Japan in the 1950s. I suggest that we look at those producing toys now to see who will emerge next!

Final Notes

Operation

One of the charming aspects of these toys is that they may all be operated by hand. They do not require an engine. Many of them are actually enhanced by hand operation since they are geared down and the faster motion produced by an engine detracts from their beauty. To operate a steam toy using an engine requires securely fastening both the toy and the engine and connecting them, usually with a spring made especially to fit between the pulleys.

Manufacturers

In the pages that follow, toys are shown by manufacturer. If a picture has two or more toys made by different manufacturers, it will be shown in the chapter about the maker of the dominant toy. Each toy, however, will be described. Where possible, toys have been pictured by the manufacturers' catalog numbers in ascending order. Some exceptions were made to provide a grouping which seemed more sensible. The manufacturers included here are:

Arnold
Becker
Bing
Carette
Doll
Eberl
Falk
Fleischmann
Günthermann
Hawkotte and Klausman
Ives
Kohnstam
Krauss, Mohr & Co.; Mohr & Krauss; Wilhelm Krauss
Märklin
Marx (Linemar)
Müller&Kadeder
Plank
Rohrseitz
Schoenner or Schönner
Weeden
Wilesco
Wunderlich
The Zschopau School

And now the symphony begins!

What follows is a short description of the toy manufacturers included in this book, and examples of the wonderful steam toys they made. Only manufacturers whose toys are shown are included. The German spelling of Nürnberg, rather than the anglicized "Nuremberg," is used throughout.

Manufacturers

Arnold

Nürnberg

The firm, founded by Karl Arnold in 1906, introduced toys with built-in flints. Arnold steam toys were produced before World War II, and are rarely marked. Perhaps the most familiar Arnold toy is the scissors sharpener, which can often be found as a wind-up with a box. The steam accessory comes in two versions. One has a vise at the end of the workbench and the other does not. Every time I have seen the former it has been mounted on a wooden base (as pictured). Many Arnold toys are misidentified and attributed to Bing, especially the sawyer, because a line drawing found in the Bing catalog resembles the Arnold toy very closely. Bing marked many of their smaller toys, however, and certainly almost all of their toys produced after 1900 (Ferris wheels and carousels excluded). I would rather err on the side of attributing any unmarked toy (as shown here) to Arnold rather than to Bing.

Arnold. Scissors sharpener, #501/1 also #S23, circa 1920. Also Falk #299/24, circa 1930 and Moko #5560. This is often misidentified as a Bing. 11.9 x 8.3 x 10.5 cm. $50-150.

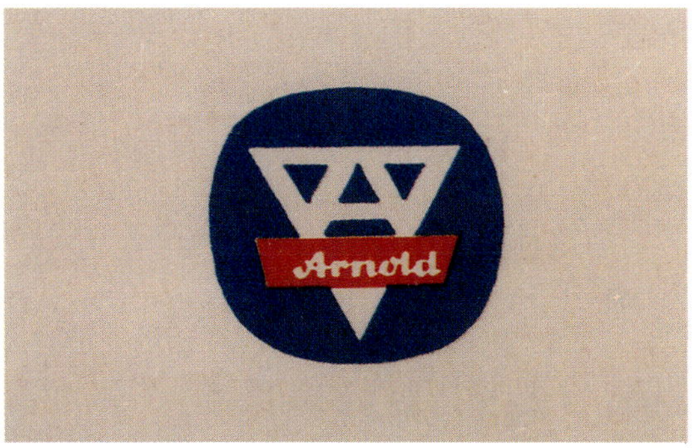

This Karl Arnold logo does not appear on steam toys.

Arnold. Sawyer with bow saw, #S23, circa 1920. Also Falk #299/24, circa 1930 and Moko #5560. 11.9 x 8.3 x 10.5 cm. $50-150.

Arnold. Sawyer at circular saw, #S23, circa 1920. Also Falk #299/24, circa 1930 and Moko #5560. 11.9 x 8.3 x 10.5 cm. $50-150.

Arnold. Sawyer at circular saw, #S23, circa 1920. This is a variation with the sawyer holding an already fashioned board (simulated, of course, from tin) instead of a round wooden log. 11.9 x 8.3 x 10.5 cm. $75-150.

Arnold. Sawyer at circular saw, circa 1910. An earlier version. Note different base and size. Heavy paint loss. 11.8 x 7 x 11.5 cm. $50-75.

Arnold. Man at stamping press, #S23, circa 1920. Also Falk #299/24, circa 1930 and Moko #5560. 11.9 x 8.3 x 11.5 cm. $100-200.

Arnold. Blacksmith at anvil, circa 1930. Moko #5131 and Hausmann #403. Hammer in right hand and brandishing iron in left. Toy is made to hold a flint. 13 x 8.4 x 13 cm. $75-150.

Arnold. Blacksmith at anvil with forge behind him, circa 1930. Moko #5553 and Hausmann #418. Hammer in right hand and brandishing iron in left. Forge has slight red glow. Toy is made to hold a flint. 17.5 x 9 x 15 cm. $300-500.

Arnold. Scissors sharpener, circa 1910. Makes sparks. Flat figure. 8.4 x 6 x 11.5 cm. $75-125.

Arnold. Scissors sharpener, circa 1905. A more elaborate version with a vise on the work table. Flat figure. I have always seen this version mounted on a wooden board. This was the first steam toy I bought. 17 x 8.2 x 14 cm. $75-150.

Arnold. Scissors sharpener, circa 1905. Also made as a windup.
Flat figure. 17 x 8.2 x 14 cm. $75-150.

Arnold. Windmill, #337, circa 1920. 10 x 10 x 18 cm. $75-150.

Arnold. Man at drill press, circa 1905. 8.5 x 6.5 x 12.5 cm. $125-200. *Provenance Gottschalk Collection.*

Arnold. Man at drill press, circa 1905. Drill press on work table. Repainted. 13.5 x 9.1 x 12.5 cm. $75-150.

Arnold. Blacksmith at anvil, circa 1905. Hammer in right hand, brandishing iron in left. 11.8 x 8.3 x 11 cm. $125-200.

Arnold. Scissors sharpener, circa 1905. 11.8 x 8.3 x 11 cm. $100-150. *Provenance Gottschalk Collection.*

Arnold. Wheelwright at anvil with stave, circa 1905. Hammer in right hand, yellow wheel at base of anvil. 14.3 x 9.7 x 15.5 cm. $250-400.

Arnold. Blacksmith at anvil, circa 1910. Hammer in right hand, brandishing iron in left. 14.2 x 8.3 x 14 cm. $300-500.

Arnold. Machine shop, circa 1910. Man at lathe, drill press to his left on work table and grinder mounted separately. Missing roof and repainted. 24 x 10.8 x 17 cm. $200-300.

Probably Arnold. Acrobat on trapeze, circa 1910. Geared down to work nicely. 12.2 x 8.5 x 23.5 cm. $300-500. *Provenance Perelman Toy Museum.*

Arnold. Blacksmith at anvil, circa 1905. Forge is at left, bellows top right. Smith working on wheel. 15.3 x 11.5 x 18 cm. $300-500.

Arnold. Blacksmith at anvil, circa 1910. Shed has bellows. Similar to Carette #108/00, circa 1906. 13.5 x 9 x 13.4 cm. $250-350. *Provenance Gottschalk Collection.*

Arnold. Workshop, circa 1910. Three-man shop. Man at left is at lathe, man at center is at drill press, and man at right is at anvil. A desirable piece. 31.7 x 15 x 24.5 cm. $800-1000. *Provenance Gottschalk Collection.*

Arnold. Wheelwright shop, circa 1910. Both wheelwrights strike wheel which revolves. Small roof covers wall. Probably scarcer than the Bing version. 17 x 8.5 x 13.5 cm. $350-500. *Provenance Gottschalk Collection.*

Arnold. Blacksmith shop, circa 1910. Blacksmith at left has hammer in right hand, brandishing iron in left. Behind him is a forge and bellows. Worker at right is filing at vise. Table has drill press to his right. 23.3 x 13.5 x 16 cm. $500-750.

Becker

See the Zschopau School.

Becker. Graduated set of three bells *(Glockenspiel)*, #7, circa 1910. 12 x 7.5 x 8 cm. $100-150.

Becker. Gnomes sawing, *(Gnomensäger)*, circa 1910. The gnomes are penny toys made by Meier. Cross-cut saw. 10.5 x 7.5 x 10.5 cm. $400-500.

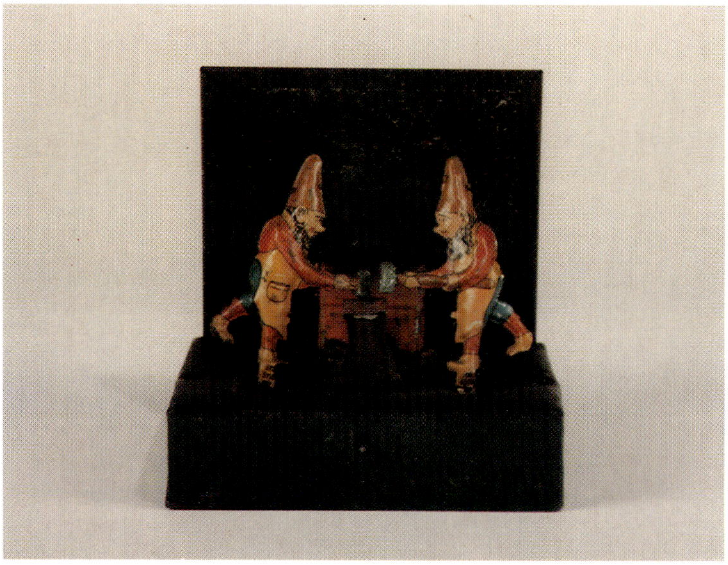

Becker. Gnomes working at anvil *(Gnomenschmiede)*, #17/12, circa 1910. The gnomes are penny toys made by Meier. 11.5 x 9.9 x 10 cm. $400-500.

Becker. Ferris wheel *(Russische Schaukel)*, #26/2, circa 1910. Replaced figures. Base is rough. 12.5 x 12.5 x 31 cm. $150-250.

Becker. Woman sawing with bow saw *(Holzsäger)*, #39, circa 1910. 12 x 7.6 x 12.5 cm. $500-750.

Becker. Man sawing with bow saw, woman cleaning branch with an axe, #39/1, circa 1910. 15 x 11.4 x 12.5 cm. $500-750.

Becker. Tool shop *(Tischlerei)*, #40/1, circa 1910. Man at left is cutting with a bow saw while the man at right is planing. 20 x 11.5 x 12.8 cm. $1200-1500.

Becker. Carousel with three horses and riders, #54, circa 1910. 9.2d x 16 cm. $200-300.

Becker. Carousel, circa 1910. Three boys sitting on platform. Balls should most likely be boys in swings as above. 8.4d x 15.5 cm. $300-500.

Becker. Carousel, #54/1, circa 1910. Two boys on platform facing each other and two boys in swings on canopy. 8.4d x 15.5 cm. $300-500.

Becker. Cooper or barrel maker *(Böttcher)*, #65, circa 1910. 8.6 x 8.6 x 11.5 cm. $150-250. *Provenance Gottschalk Collection.*

Becker. Chimney sweep standing on chimney *(Essenkehrer)*, #74, circa 1910. 8.6 x 8.6 x 15.5 cm. $350-500.

Becker. Man at circular saw, #80, circa 1910. 11 x 8.5 x 11.5 cm. $100-200. *Photograph by Françoise R. Hirschberg.*

Becker. Chimney sweep in chimney *(Essenkehrer in Schornstein)*, #75, circa 1910. Sweep lowers and rises from chimney as he rotates. 9.4 x 8.6 x 10 cm. $350-500. *Photograph by Françoise R. Hirschberg.*

Becker. Miner *(Bergmann)*, #83, circa 1910. 12.8 x 10 x 11.5 cm. $350-500.

Becker. Miners *(Bergwerk),* #83/2, circa 1910. Man at left is hammering ore while man at right is crushing ore with trip hammers. 24.2 x 10.6 x 17.6 cm. $500-750.

Becker. Snare drummer *(Tambour),* #90, circa 1910. 8.6 x 8.6 x 11.5 cm. $300-500.

Becker. Sausage factory *(Wurstfabrik),* #92, circa 1910. Man at left is cleaving dog's head while man at right is grinding meat. 22.6 x 11 x 17.2 cm. $1200-1500.

Becker. Miners, #83/4, circa 1910. Miner kneeling at left has pick while standing miner at right has hammer. 21.5 x 10.5 x 11 cm. $500-750.

Becker. Butcher cleaving dog's head *(Fleisher)*, #93/1, circa 1910. 12.3 x 8 x 12.5 cm. $600-800.

Becker. Mason, circa 1910. Man holds trowel in right hand and brick in left. Smokes a pipe. 12.3 x 8 x 12.5 cm. $500-750.

Becker. Wine maker, circa 1910. Man is pumping wine from the barrel into the bucket. 13 x 9.2 x 13 cm. $400-600.

Becker. Workman with hammer, circa 1910. Man holds hammer in right hand and strikes vise. Tools on work table. 12.3 x 8 x 13 cm. $400-600.

Becker. Man sharpening axe, circa 1910. Toys similar to this have been identified as Mohr & Krauss, circa 1905. 11.2 x 7.5 x 12 cm. $75-150.

Becker. Detail of tools on work table. Hammer, pump, and rasp.

Becker. Soldiers at guard house *(Wachtparade),* #105, circa 1910. Three soldiers revolve while the fourth stands guard at sentry station. Replaced flag. Paint loss on sentry station. 19.4 x 14.4 x 18.5 cm. $600-800. *Photograph by Françoise R. Hirschberg.*

Becker. Clown striking bells *(Musikalischer Clown),* #174/1, circa 1910. 13 x 9 x 16.5 cm. $500-750.

Becker. Bear tamer, #189, circa 1910. Girls reading form background. Paper lithography. Bear tamer has switch in right hand. Bear goes through hoop. Plays musical notes. 16.3 x 8.9 x 12.3 cm. $500-750.

Becker. Chimney sweep fending off dog, #179, circa 1910. Outdoor scene with tree in background. 15.5 x 9.7 x 12.8 cm. $500-750.

Becker. Card players *(Kartenspieler),* #224, circa 1910. 16.5 x 6.7 x 11.5 cm. $1000-1500.

Becker. Fighting roosters *(Hahnkampf),* #252K, circa 1910. Roosters are penny toys by Meier. 12.1 x 7.5 x 11.5 cm. $500-750.

Becker. Miller with windmill and mule, #257, circa 1910. Mill has trip hammers just visible on left side. Miller has switch in right hand. 18 x 18.5 x 24 cm. $1300-1500.

Becker. Clown with dogs in umbrella, #501, circa 1910. Umbrella twirls. 14 x 7.3 x 17 cm. $1500-2000.

Becker. Man at drill press, circa 1910. 9.9 x 6.9 x 12 cm. $125-200.

Becker. Man at press, circa 1910. Incomplete. 10 x 7 x 11 cm. $125-200. *Provenance Gottschalk Collection.*

Becker. Man at grindstone, circa 1910. 9.8 x 6.8 x 11.5 cm. $100-150.

Becker. Man at lathe, circa 1910. Long work bench missing mechanism, probably a drill press. 17 x 9.2 x 11.5 cm. $150-250.

Becker. Iron factory, circa 1910. Man at left is hammering at anvil while man in center is working forge and man at right is smelting with a furnace behind him. Separate roofs. 27.5 x 13.5 x 21 cm. $1000-1500.

Becker. Human carousel, circa 1910. Three men hanging by their arms from canopy. 11d x 20 cm. $500-750.

Becker. Wrestlers, circa 1910. 11.2 x 5.5 x 13.5 cm. $750-1000.

Becker. Violinist performing for student, circa 1910. Student is a penny toy by Meier. 12.2 x 7 x 11.2 cm. $750-1000.

Becker. Girl skipping rope, #178, circa 1910. 15 x 8.4 x 21 cm. $250-500. *Provenance Gottschalk Collection.*

Becker. Tennis players, circa 1910. 18.6 x 7.8 x 12.5 cm. $1000-1500.

Becker. Harpist, circa 1910. Background theater curtains. 12 x 9.5 x 12.5 cm. $1250-1500.

Becker. Blacksmith shop, circa 1910. Smith is working forge which has bellows. Vise on work table. 21 x 9 x 15 cm. $600-800. *Provenance Perelman Toy Museum.*

Becker. Dredge, circa 1910. Woman at left is at water pump filling pail while man at right is working circular saw. Really a hand-cranked toy. Also pictured in Ullmann and Engelmann catalogs. 24 x 19 x 28 cm. $500-750.

Becker. Blacksmith shop, circa 1910. Smith at left is working bellows while smith at right is hammering at anvil. 23.6 x 12.6 x 19 cm. $1000-1500. *Provenance Gottschalk Collection.*

Bing

Nürnberg

Established in either 1863 or 1866 as a wholesale business selling toys under the name Gebrüder Bing, this company started its own toy production in 1879. Bing produced hundreds of different steam toys, but (as we shall see when we talk about the Zschopau School) many were produced by other manufacturers for them. Bing steam accessories are highly desirable. They are colorful and well-articulated. I cannot find references earlier than 1898 and toys of earlier vintage bear no trademark. Bing used a number of trademarks, which were embossed on the tin-plate, painted on, or stamped on a separate metal plate and affixed to the toy. Few Ferris wheels or carousels bear a trademark. Bing at one time employed thousands of workers. In the early 1930s Bing fell on hard times, and Bub, Falk, and Krauss assumed production of many of the steam toys.

A Bing mark used until circa 1902, so far not seen on toys.

A Bing mark used 1902-1907, not seen on toys.

A Bing mark used 1908-1925.

A variation of the Bing mark used 1908-1925.

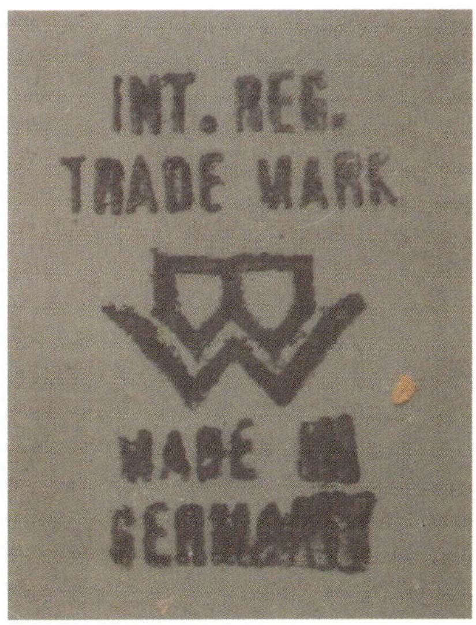

A Bing mark used 1925-1932.

A Bing mark used circa 1900.

Bing. Sawyer with bow saw *(Holzsäger),* #10/26, circa 1927. 15.8 x 6.8 x 13.5 cm. $150-300.

Bing. Blacksmith *(Schmied),* #10/28, circa 1927. Hammer in right hand, brandishing iron in left. 13.3 x 6.6 x 13.5 cm. $200-300.

Bing. Butcher *(Fleischer),* #10/27, circa 1927. 13.3 x 6.9 x 13.5 cm. $300-500.

Bing. Man at circular saw *(Holzsäger),* #10/231, circa 1927. Sawyer cuts board in half. 16 x 7.7 x 13.5 cm. $250-500.

Bing. Man at dredge *(Schoprad),* #10/243, circa 1927. 16 x 7.7 x 13.5 cm. $250-500.

Bing. Man sifting *(Sandsieb)*, #10/244, circa 1927. 15.8 x 7.7 x 13.5 cm. $250-500.

Bing. Organ grinder with music *(Orgelspieler mit Musikwerk)*, #10/291 or #10/292, circa 1927. The music box is ornately lithographed with a dancing couple under a set of gold-painted pipes. 10/291 has 22 notes and 10/292 has 8 notes. 13.3 x 6.9 x 13.5 cm. $800-1200.

Bing. Man at pump *(Pumpbrunnen)*, #10/245, circa 1927. 15.8 x 7.7 x 13.5 cm. $250-500.

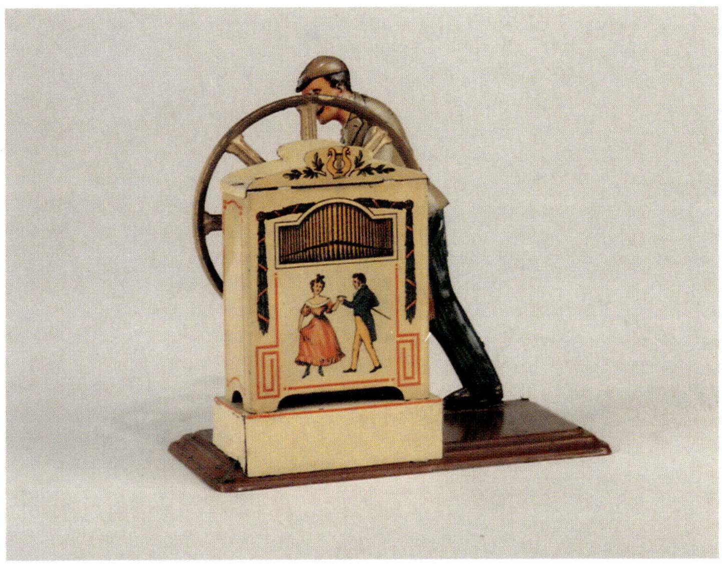

Bing. Detail on the reverse side of organ grinder.

Bing. Sawyers *(Holzhauer und Säger)*, #10/247, circa 1930. Man on left is chopping wood with an axe, while man at right is using a bow saw. 25 x 9 x 15 cm. $500-750.

Bing. Blacksmiths working at anvil *(Schmiede)*, #9956/355, circa 1912. Blacksmith at left has hammer in right hand, brandishing iron in left. Smith at right has hammer in both hands. 19.5 x 8.5 x 13 cm. $300-500.

Bing. Wheelwrights *(Wagenschmiede)*, #9956/358, circa 1912. Each smith has a hammer in his hand. Wheel rotates as they work on it. 19.5 x 8.5 x 13.5 cm. $500-750.

Bing. Sawyers with cross-cut saw *(Zimmerleute Sägemänner)*, #10/246, circa 1930. 25 x 9 x 15 cm. $500-750.

Bing. Windmill *(Windmühle)*, #10/21, circa 1930, also #9956/228, circa 1912. The smallest of the Bing windmills. 7.4d x 17 cm. $50-100.

Bing. (At left) A small windmill, #9956/254, circa 1907. 6.8d x 12 cm. $25-50. (At right) A windmill driving a grinder, circa 1907. 17.5 x 7.5 x 10 cm. $50-100.

Bing. Windmill with two trip hammers, circa 1907. 17.5 x 8.5 x 18 cm. $100-200. *Photograph by Françoise R. Hirschberg.*

Bing. Windmill with music *(Windmühle mit Musik),* #10/241, circa 1930, also #9956/381, circa 1912. 12 x 12 x 22.6 cm. $150-250.

Bing. Windmill *(Windmühle),* #9956/229, circa 1912. 12 x 12 x 21 cm. $200-300. *Photograph by Françoise R. Hirschberg.*

Bing. Windmill *(Windmühle),* #10/2115, circa 1927. 17.2 x 10.8 x 21 cm. $150-250.

Probably Bing. Windmill with trip hammers *(Windmühle)*, circa 1930. 12.9 x 12.9 x 30.8 cm. $250-350. *Photograph by Françoise R. Hirschberg.*

Bing. Mills *(Windmotor)*, #10/248, circa 1927. House at left, windmill in center, mill with trip hammers at right. 25.6 x 15 x 25.5 cm. $300-500. *Photograph by Françoise R. Hirschberg.*

Bing. Windmill with miller, #10/280, circa 1930. Miller goes in and out of mill, reversing direction while he does so. Trip hammers to left under shed. Toy is surprisingly heavy. 19 x 16 x 26 cm. $800-1200.

Bing. Dredge *(Baggerwerke)*, #10/281, circa 1932. Dredge has five buckets. 21 x 9 x 16.5 cm. $400-600.

Bing. Drill press, #10/290, circa 1927. Moveable table. 13.4 x 6.8 x 16 cm. $250-350.

Bing. Shoemaker's apprentice *(Schusterlehrling)*, #9956/475, also #10/228, circa 1912. 11.2 x 7 x 12 cm. $400-600. *Provenance Gottschalk Collection.*

Bing. Master tailor *(Schneidermeister)*, #9956/474, also #10/227, circa 1912. 11.2 x 7 x 12 cm. $300-500.

Bing. Master shoemaker *(Schuhmakermeister)*, #9956/476, also #10/229, circa 1912. 11.2 x 7 x 12 cm. $300-500.

Bing. Master tinsmith *(Flashermeister)*, #9956/477, also #10/230, circa 1912. 11.2 x 7.1 x 12 cm. $300-400.

Bing. Ferris wheel *(Russische Schaukel)*, #9956/239, circa 1912. Russian swings. Four original porcelain women figures. 13 x 10 x 30 cm. $500-750.

Bing. Ferris wheel, #10/299, circa 1930. Eight brightly lithographed gondolas. Wheel may have been repainted. 24.5 x 11.6 x 27.6 cm. $750-1250. *Provenance Gottschalk Collection.*

Bing. Carousel, #9956/416, circa 1912. Four riders on horseback. Body of carousel repainted. 21d x 34.3 cm. $1200-1500. *Photograph by Françoise R. Hirschberg.*

Bing. Carousel, #10/295 and #10/296, circa 1930. Riders are on horse, ostrich, elephant, and in car. Wonderful condition. 17.4d x 30.8 cm. $1500-2000.

Probably Bing. Carousel, circa 1930. Four riders on horseback.
22.6d x 27.9 cm. $1500-2000.

Probably Bing. Carousel, circa 1930. Riders in scalloped seats. Three dirigibles hang from canopy. A rare piece. Replaced figures. 18.8d x 33.6 cm. $3000-4000.

Bing. Violinist *(Violinspieler),* #10/2101, circa 1930. Uncharacteristically crude for a Bing. 11.2 x 7 x 12 cm. $800-1000.

Bing. Clown with poodle, #8743/9, circa 1902. One in a series of twelve. 11.3 x 7.2 x 11.5 cm. $500-750.

Bing. Chauffeur putting air in tire, #10/2102, circa 1927. Art Deco. 11.3 x 7.5 x 12.5 cm. $1000-1500.

Bing. Girl watering flowers, #8743/3, circa 1902. One in a series of twelve. 11.3 x 7.2 x 11.5 cm. $1250-1750. *Photograph by Françoise R. Hirschberg.*

Bing. Shoe shiner, #8743/5, circa 1902. One in a series of twelve. 11.3 x 7.2 x 11.5 cm. $1250-1750. *Photograph by Françoise R. Hirschberg.*

Bing. Woman feeding geese and rooster, #8743/10, circa 1902. One in a series of twelve. 11.3 x 7.2 x 11.5 cm. $1250-1750.

Bing. Musician with music, #8743/11, circa 1902. One in a series of twelve. 11.3 x 7.2 x 11.5 cm. $1250-1750.

Bing. Organ grinder with music, #8743/12, circa 1902. One in a series of twelve. 11.3 x 7.2 x 11.5 cm. $1250-1750.

Bing. Polisher *(Polirer)*, #9956/23, circa 1902. 11.3 x 7.2 x 14 cm. $400-600.

Bing. Sawyer with bow saw *(Holzsäger)*, #9956/347, circa 1930, also #10/26, circa 1912. 16 x 6.8 x 14 cm. $100-200.

Bing. Grinder *(Schleifer),* #9956/25, circa 1902. 10.3 x 6.8 x 14 cm. $400-600. *Photograph by Françoise R. Hirschberg.*

Bing. Locksmith *(Schlosser),* #9956/99, circa 1902. 13.5 x 7 x 14 cm. $400-600.

Bing. Freight elevator *(Waaren-Aufzug),* #9956/51, circa 1898. Repainted. 14.4 x 17.2 x 42.5 cm. $100-200. *Photograph by Françoise R. Hirschberg.*

Bing. Sawyer at circular saw *(Holzsäger),* #9956/116, circa 1902. Sawyer is cutting wooden board in two. 15 x 7.5 x 14 cm. $400-600.

Bing. Smith at drop press *(Schmiedepresse),* #9956/359, circa 1912. 16 x 7.5 x 21.5 cm. $500-750.

Bing. Organ grinder *(Orgelspieler)*, #9956/275 and #9956/495, circa 1912. Twenty-two notes and eight notes. 13.2 x 7 x 12.7 cm. $1000-1250. *Photograph by Françoise R. Hirschberg.*

Bing. Man planing and sawyer *(Schreiner–Säger und Hobler)*, #9956/368, circa 1912. Planer, left, has horned plane; sawyer, right, has cross cut saw. Vises on work table. Notice different style figures. 15.5 x 15.5 x 16.5 cm. $750-1000. *Photograph by Françoise R. Hirschberg.*

Bing. Straw cutters *(Häckselschneider)*, #9956/366, circa 1912. Man on left is cutting while man on right is providing power by turning wheel. 18 x 13 x 18 cm. $750-1000.

Bing. Blacksmith shop *(Schmiede)*, #9956/167, circa 1912. Smith at left is working bellows while smith at right is hammering a wheel on the anvil. 21 x 11 x 17 cm. $1000-1500. *Provenance Gottschalk Collection.*

Bing. Butter churn, circa 1906. 9.8 x 6.7 x 9 cm. $25-50.

Bing. Turnip cutter, #9956/242, circa 1906. 11.5 x 8.6 x 12 cm. $150-300.

Bing. Man at drill press, #9956/243, circa 1906. 11.3 x 8.5 x 14.5 cm. $150-300.

Bing. Butter maker, #9956/246, circa 1906. 11.5 x 8.8 x 12 cm. $150-300.

Bing. Sawyer at circular saw *(Säger)*, #9956/244, circa 1906. 11 x 7.3 x 11.5 cm. $150-300.

Bing. Hat presser, circa 1906. 11.2 x 8.2 x 11.5 cm. $150-300.

Bing. Launderer, #9956/245, circa 1906. 11.2 x 8.5 x 12 cm. $150-300.

Bing. Mason, circa 1906. 11.2 x 8.4 x 12 cm. $150-300.

Bing. Butcher *(Schlachter),* circa 1906. Cutters in both hands. 11.5 x 7.2 x 12 cm. $150-300.

Bing. Smith working at rock, circa 1906. Hammer in right hand, brandishing iron in left. 11.5 x 7.2 x 12 cm. $150-300.

Bing. Wheelwright, circa 1906. Hammer in right hand, wheel in left. 11.4 x 7.2 x 11.5 cm. $150-300.

Bing. Butcher *(Schlachter),* circa 1906. Cutter in right hand, bone in left. 11.5 x 7.2 x 12 cm. $150-300.

Bing. Sorting machine, #9956/62, circa 1902. 25 x 11.8 x 17 cm. $500-750. *Photograph by Françoise R. Hirschberg.*

Bing. Grinding machine, #9956/134, circa 1906. 13.3 x 7 x 15.6 cm. $350-500.

Bing. Thrasher, circa 1906. 16.4 x 10.5 x 15 cm. $400-600.

Bing. Fountain, #9956/489, circa 1925. 14.8d x 11 cm. $50-100.

Bing. Fountain with female figure, #9956/80, circa 1906. Carette and Schoenner made similar fountains. 21.5d x 20 cm. $300-500.

Bing. Sifter or sand toy, #9956/166, circa 1904. Missing funnel. Also made in a smaller version as #9956/165. 17.5 x 9.6 x 10.5 cm. $50-100.

Bing. Dredge, #9956/180, circa 1906. Notice the round tower, unlike the square tower shown in the Bing catalogs. 17.2 x 7.6 x 18.5 cm. $150-250.

Bing. Right-angled transmission *(Winkel transmission),* #9956/345, circa 1912. 15.5 x 13.4 x 8 cm. $25-50.

Bing. Drill press, #9956/139, circa 1900. 6.5d x 11.8 cm. $50-100.

Bing. (At left) Washing machine, circa 1902. 11.5 x 8 x 7.3 cm. $125-200. (At right) Early sewing machine, circa 1902. 11.6 x 6 x 6 cm. $50-100.

Bing. Frame saw, #9956/261, circa 1907. 6.4 x 9.4 x 12.5 cm. $50-100.

Bing. Sawmill, circa 1906. 10.6 x 10.4 x 15 cm. $100-200.

Bing. Exhaust fan, #9956/202, circa 1906. 12.7 x 10.8 x 11.4 cm $300-500. *Photograph by Françoise R. Hirschberg.*

Bing. Water mill with trip hammers, circa 1930. 18 x 15.2 x 14 cm. $150-250.

Bing. Water mill, circa 1930. Notice bridge on right. Roof has been resoldered. 19 x 18 x 19.5 cm. $100-150.

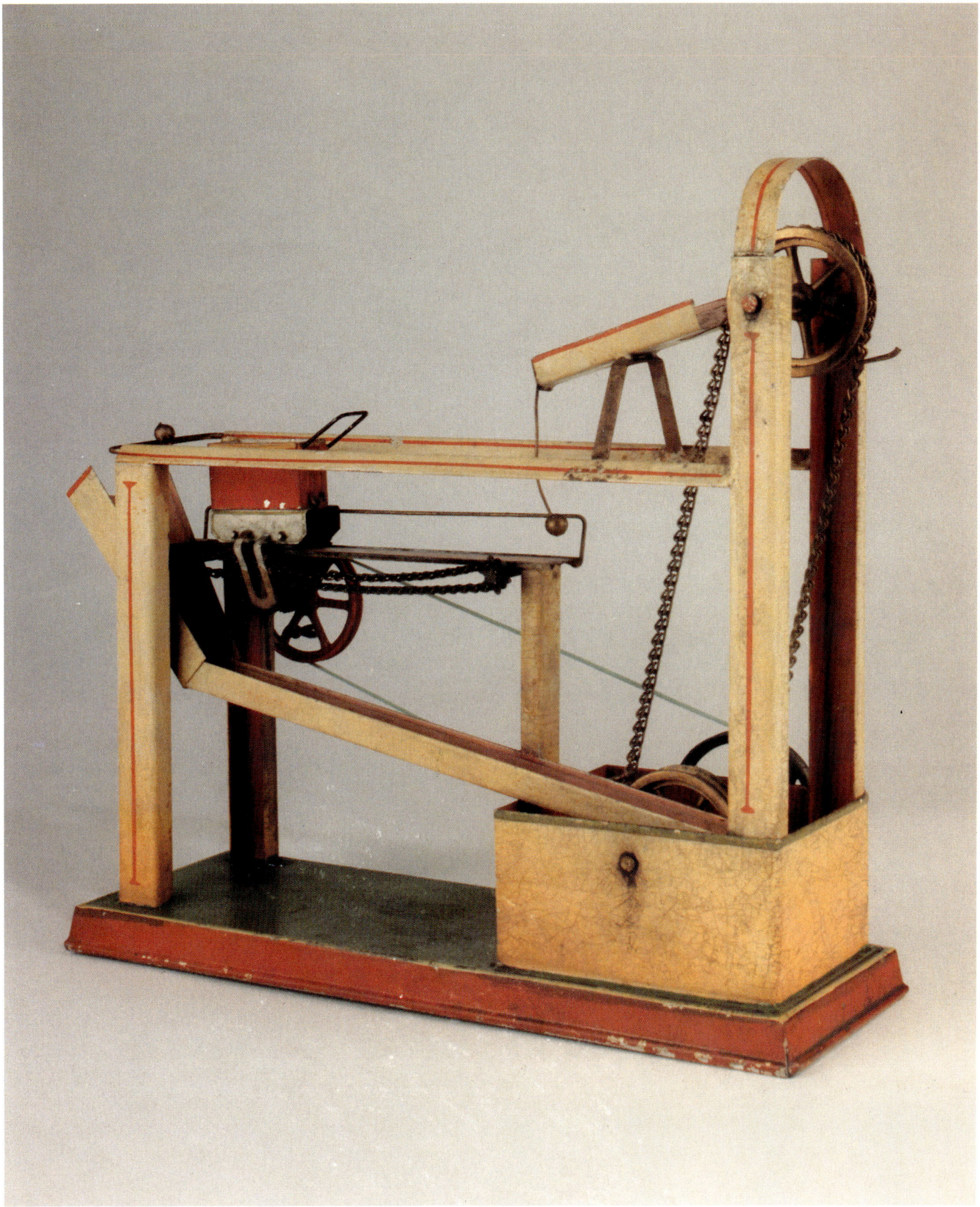

Bing. Marble mill with cart, circa 1930. Cart tilts and deposits marble in reservoir. 29.5 x 11.4 x 29 cm. $500-750.

Bing. Water mill with trip hammer, #9956/319, circa 1912. 14 x 12 x 10.5 cm. $50-100.

Bing. Water mill with trip hammer *(Hammerwerk mit Wassermühle)*, #9956/305, circa 1912. 18.3 x 13.3 x 13 cm. $100-200. *Photograph by Françoise R. Hirschberg.*

Bing. Six-sided fountain, circa 1930. 20d x 8 cm. $100-200.

Bing. Linen mill, circa 1890. Rare. Wonderful action as weaving is simulated. Original condition. 26 x 15.2 x 20 cm. $1250-1750. *Photograph by Françoise R. Hirschberg.*

Bing. Cotton mill, circa 1890. Very rare. A wonderful toy with loom on left, linen mill in center, and washer on right. Notice the linen mill shown previously forms the center piece of this toy. 63.5 x 18.5 x 36.8 cm. $4500-6000. *Photograph by Françoise R. Hirschberg.*

TTON MILL.

Bing. Coal mine *(Bergwerk)*, #9956/289, circa 1912. At left is dredge with bell at top of tower. In the center are men moving coal carts, one not visible. At right is a steam plant. Notice two miners with picks working in the mine. 40.6 x 21 x 21 cm. $2000-3000.

Bing. Dog with drum playing cymbals, circa 1900. 11.5 x 7.2 x 17 cm. $750-1000. *Provenance Perelman Toy Museum.*

Bing. Lighthouse *(Leuchtturm)*, circa 1906. Partial repaint. 10.3 x 10.3 x 29 cm. $1750-2500.

Bing. Lighthouse *(Leuchtturm),* circa 1898. Small boat rotates in moat around lighthouse. 21d x 25.5 (with sheds) x 31.8 cm. $1000-1500.

Bing. Tailor shop *(Schneiderwerkstätte),* #9956/378, circa 1912. Triple action: both tailors move, as does the woman's head in the window. Left tailor is sewing, right tailor ironing. One of a set of three toys. 22 x 8 x 16 cm. $1000-1500. *Provenance Atlanta Toy Museum.*

Bing. Tinsmith shop *(Klempnerwerkstätte),* #9956/380, circa 1912. Both figures move. Smelting pot centered on table. One of a set of three toys. 22 x 8 x 16 cm. $1000-1500. *Provenance Hoffman Collection.*

Bing. Tailor and shoe repair shop, ca.1900. Early, hand-painted and lithographed. Windows open, door on side opens, and roof comes off. Probably an early prototype for later toys. Rare. 21.5 x 8 x 16.5 cm. $1500-2000.

Bing. Interior view of preceding toy with roof off.

Probably Bing. Boxer punching bag, circa 1900. Four spheres simulate punching bag. 10.5 x 7.3 x 14.5 cm. $400-600.

Probably Bing. Blacksmiths at anvil, circa 1890. Both smiths are hammering at anvil. Smith at right has brandishing iron in left hand. Could possibly be Carette 10.8 x 6.5 x 13 cm. $400-600. *Photograph by Françoise R. Hirschberg.*

Probably Bing. Blacksmiths, circa 1890. Smith at left is at forge and works bellows. Smith at right is hammering. Could possibly be Carette. 15 x 6 x 13 cm. $200-350. *Photograph by Françoise R. Hirschberg.*

Probably Bing. Miller with mill and mule, circa 1890. Miller has switch in right hand to whip mule. Could possibly be Carette. Old repaint. 12 x 7.5 x 19.5 cm. $300-500.

Probably Bing. Factory, circa 1900. Four machines. (Back row) Vertical saw on left, press on right. (Front row) Circular saw on left, grinder on right. Some repainting. 30.5 x 25 x 35.5 cm. $500-750.

Bing. Monkey turning wheel, circa 1890. As the monkey turns the wheel, a lead weight strikes the bell, causing it to ring. 17 x 10 x 17 cm. $1000-1500.

Probably Bing. Musicians, circa 1890. Man at left is playing bass while man at right is playing violin. Could possibly be Carette. 17.8 x 5.8 x 14.5 cm. $800-1000. *Photograph by Françoise R. Hirschberg.*

Bub

Nürnberg

The characteristically recognizable steam toys once attributed to Karl Bub are now attributed to Heino Becker and the Zschopau School. I, too, am going to attribute these point-nosed, near square-capped rarities to the Zschopau School. The Bub catalog reprinted in Die Anderen Nürnberger shows no steam toys. It is now believed that Bub was a retailer for steam toys. Although Bub assumed a large part of the Bing company, I have not found any steam toys with any of the Bub marks. Karl Bub founded the company that bears his name in 1851. The company remained in business until the mid-1960s.

Carette

Nürnberg

George Carette et Cie was founded in 1886 by native Frenchman George Carette. Carette was associated with Bing, and acted as a supplier for them. The ties between Bing, Carette, and Becker (and perhaps other Zschopau producers) may never be fully known. The Carette factory closed in 1917. These toys are highly desirable, especially the early toys made at the turn of the century, such as the organ-grinder and man at the printing press.

A Carette mark used 1905-1917, not seen on toys.

Carette. Miner, #195/2, circa 1911. Miner chipping at wall with pick. 11 x 7.5 x 14 cm. $300-400. *Provenance Gottschalk Collection.*

Carette. Man planing, #195/5, circa 1911. Uses horn plane. 11 x 7.5 x 13.5 cm. $300-400.

Carette. Man at circular saw, #195/6, circa 1911. Cutting wooden board in half. 11 x 7.5 x 13.5 cm. $300-400.

Carette. Man at drill press *(Bohrer),* #195/7, circa 1911. 9.8 x 9.8 x 13.5 cm. $350-500.

Carette. Scissors sharpener, #195/8, circa 1911. 11 x 7.5 x 13.5 cm. $300-400.

Carette. Two blacksmiths at anvil, #195/11, circa 1911. Each smith has a hammer. 18.2 x 8.5 x 14 cm. $350-500.

Carette. Sawyer with bow saw *(Holzsäger)*, #195/14, circa 1911. 11 x 7.5 x 14 cm. $300-400.

Carette. Butcher, #195/16, circa 1911. Three knives on butcher block. 14 x 7.5 x 13.5 cm. $400-600.

Carette. Woodsman with axe, #195/15, circa 1911. Chopping square block. 11 x 7.5 x 14 cm. $300-400.

Carette. Man at lathe, #195/17, circa 1911. 9.8 x 9.8 x 13.5 cm. $400-600. *Provenance Gottschalk Collection.*

Probably Carette. Man at circular saw *(Holzsäger)*, circa 1900. 14 x 6.9 x 12.5 cm. $75-150.

Probably Carette. Wheelwright with stave, circa 1900. 11.8 x 7 x 12.5 cm. $250-400.

Carette. Man at drill press *(Bohrer)*, #698/16, circa 1905. Vise on work table. Possibly missing second figure. 16.5 x 8.5 x 13 cm. $200-300.

Carette. Workshop, 698 series, circa 1905. Man on left at circular saw, right and center figures using horn planes. 22.5 x 10 x 13 cm. $300-500. *Provenance Gottschalk Collection.*

Probably Carette. Sawmill, circa 1902. 22.8 x 9.8 x 19.3 cm. $300-500.

Carette. Thrashing machine, #614/9, circa 1911. 26 x 10 x 14.5 cm. $400-600. *Photograph by Françoise R. Hirschberg.*

Probably Carette. Lathe *(Drehbank),* circa 1911. Tool position adjustable. 14 x 7.4 x 10 cm. $75-150.

Carette. Fountain with pump, circa 1911. 11.1d x 21.5 cm. $75-150.

Carette. Fountain, #198/0, circa 1902. 16.9 x 10.6 x 16.5 cm. $150-250.

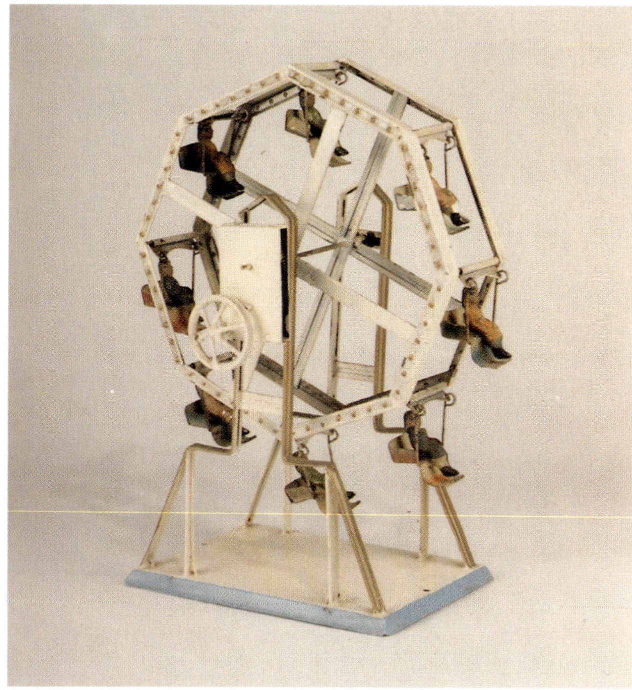

Probably Carette. Ferris wheel *(Russische Schaukel),* circa 1900. Russian swings. Eight seats. 16.6 x 13 x 28 cm. $500-750.

Probably Carette. Carousel, circa 1911. Two boys on swings and two on horseback. 19d x 23 cm. $750-1000.

Carette. Carousel, #138, circa 1911. Double canopy. Three children in swings hanging from top canopy. Three riders on horseback. Also pictured in Ullmann & Engelmann catalogs. 15d x 24 cm. $1000-1500.

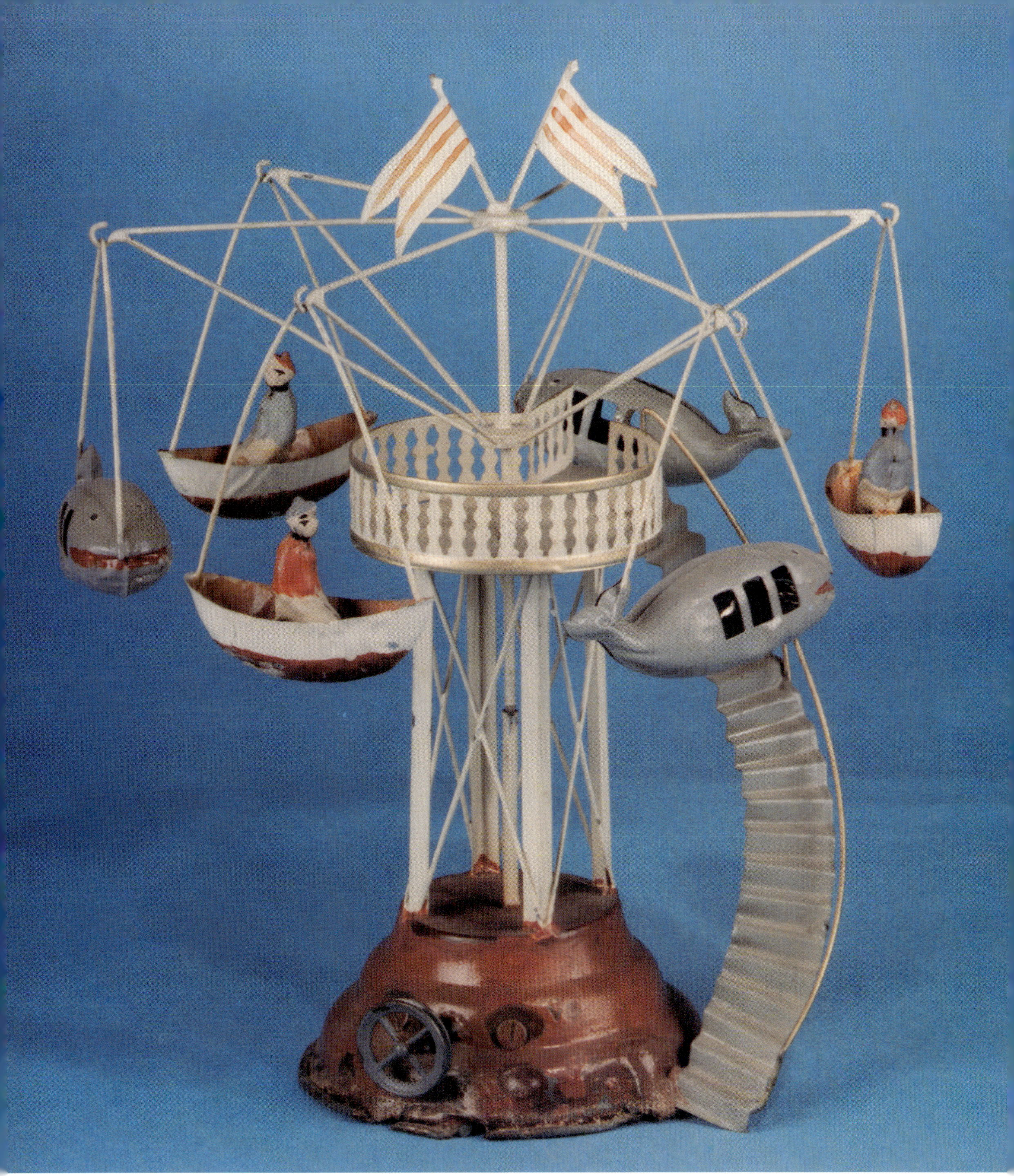

Carette. Carousel, #138/5, circa 1905. Three boys in boats, men in whales, circular staircase. My favorite toy. 20d x 25.5 cm. $2000-3000.

Carette. Grinder *(Schleifer),* #473S, circa 1902. 11 x 7.2 x 15 cm. $300-400.

Carette. Man at lathe *(Drechsler),* #615D, circa 1902. 11 x 9.6 x 14 cm. $400-600. *Photograph by Françoise R. Hirschberg.*

Carette. Factory *(Schlosserei),* #626/20, circa 1911. Five figures: grinder, drill press operator, filer, smith at anvil, and smith at forge with bellows. 42 x 16 x 22 cm. $4000-6000. *Provenance Perelman Toy Museum.*

Carette. Polisher *(Polirer),* #625P, circa 1902. 9 x 7.3 x 15 cm. $300-400.

Carette. Dredge with Archimedes Screw, #631, circa 1902. Heavy paint loss. 23 x 11.8 x 17.5 cm. $100-150.

Carette. Windmill with dredger, #697/14-1/2, circa 1911. 25 x 11.2 x 19 cm. $125-200.

Carette. Dredge *(Baggerwerk)*, #631/2, circa 1902. Dredge has four buckets. Also made by others. 25 x 11 x 25 cm. $300-500.

Carette. Water elevator *(Grosses Schöpfwerk)*, #631/6, circa 1911. 42.5 x 24.2 x 25.4 cm. $750-1500.

Carette. Dredger *(Backerwerk mit Roll-Umkippwagen)*, #631/4, circa 1911. Wagon rolls and tips. 42 x 11.2 x 30.8 cm. $1500-2000.

Carette. Windmill with dredge and water wheel, #631/18-1/2, circa 1911. Ullmann & Engelmann #2207. Often attributed to Schoenner. Mill is missing sails. These always seem to look beaten up. 18.5 x 21.5 x 24.5 cm. $150-250.

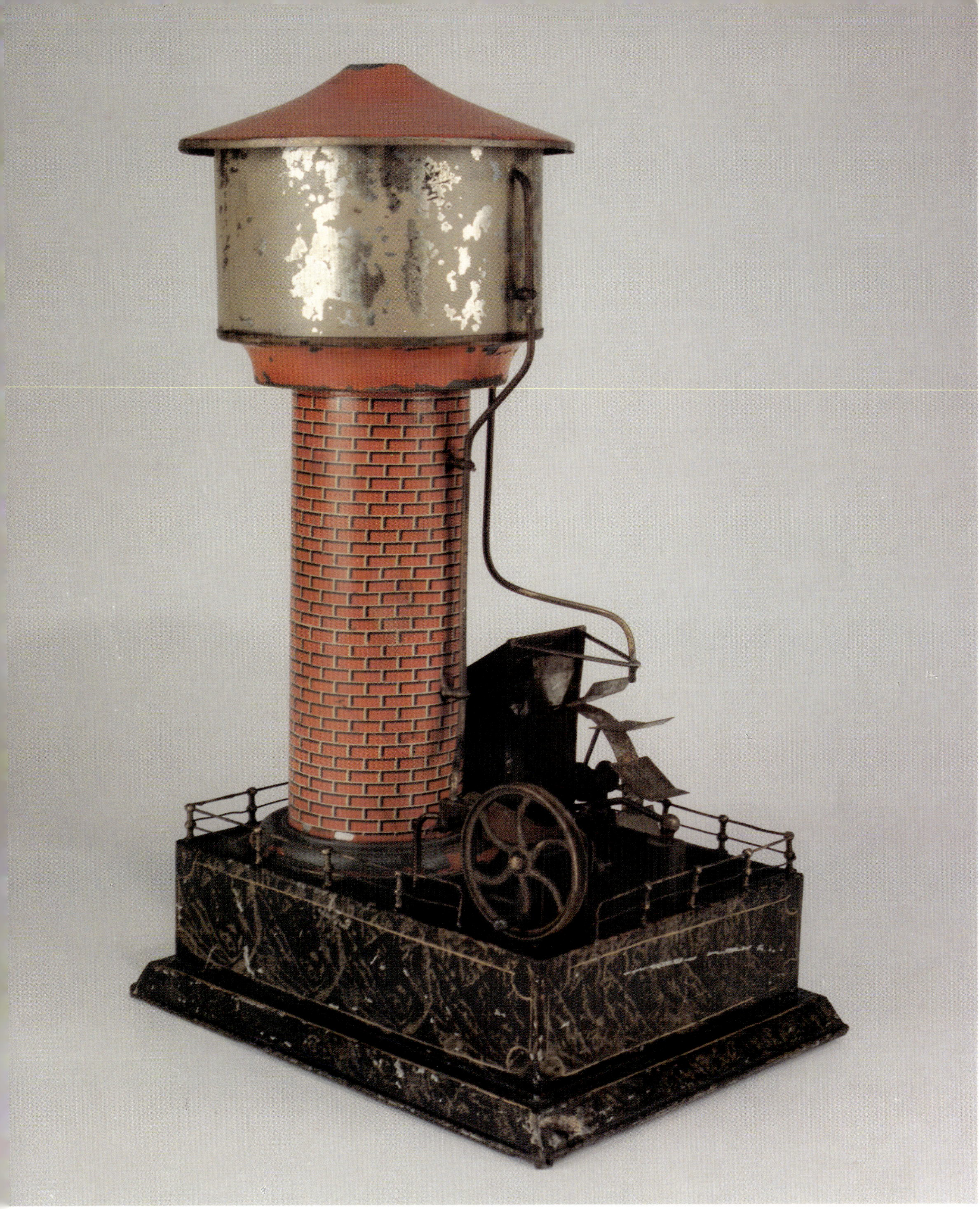

Carette. Water tower, circa 1920. Has reservoir and water wheel.
Paint loss on top of tower. 23.8 x 19 x 39.2 cm. $1500-2000.

Carette. Man at printing press *(Druckpresse mit Beweglicher Figur)*, #634D, circa 1902. Rubber roller makes farm scene. Wooden base with paper lithography. 13 x 12.8 x 13 cm. $600-800. *Photograph by Françoise R. Hirschberg.*

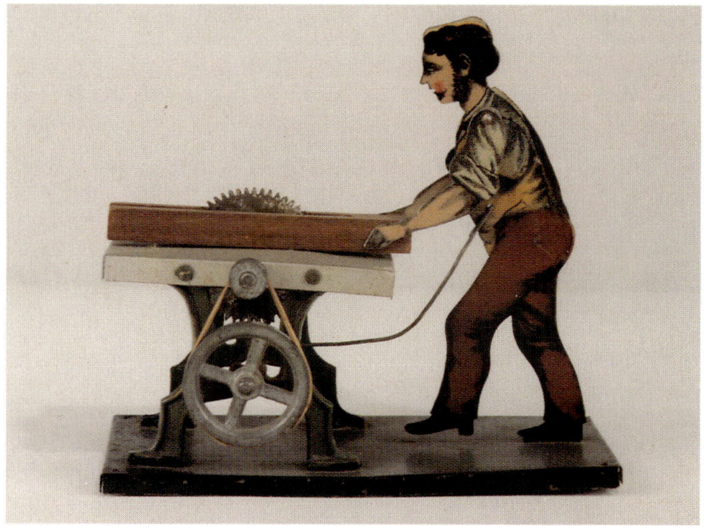

Carette. Sawyer at circular saw *(Holzsäger)*, circa 1902. Cuts wooden board in half. 18.3 x 7.9 x 13.5 cm. $450-600.

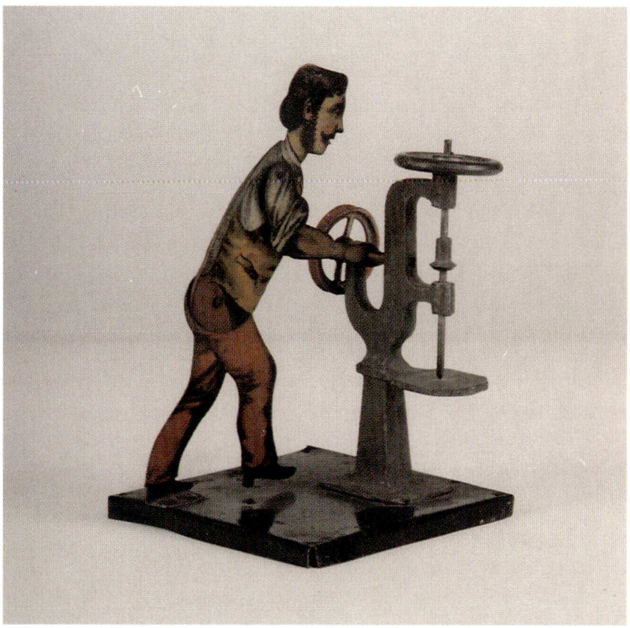

Carette. Man at drill press *(Bohrer)*, circa 1902. 8 x 8 x 14 cm. $200-300. *Provenance Gottschalk Collection.*

Carette. Organ grinder *(Orgelspieler)*, #657, circa 1902. Various versions play six, twenty-four, and forty-eight notes. 11.7 x 7.2 x 13 cm. $750-1000. *Provenance Gottschalk Collection.*

Carette. Reverse side of the organ grinder #657, showing lithography of organ.

Carette. Bootblack *(Stiefelputzer),* #678/00, circa 1902. 11.2 x 7 x 13 cm. $750-1000.

Carette. Bootblack with Dandy *(Stiefelputzer mit Gigerl),* #678/0, circa 1902. Dandy holds walking stick. Bootblack is same figure as on previous toy, #678/00. 16 x 8 x 16.5. $1000-1500. *Provenance Gottschalk Collection.*

Carette. Acrobatic clowns *(Clowns mit Doppelrack)*, #682/1, circa 1902. Also made by Bing as 9956/36. 23.7 x 10.5 x 19.5 cm. $300-500.

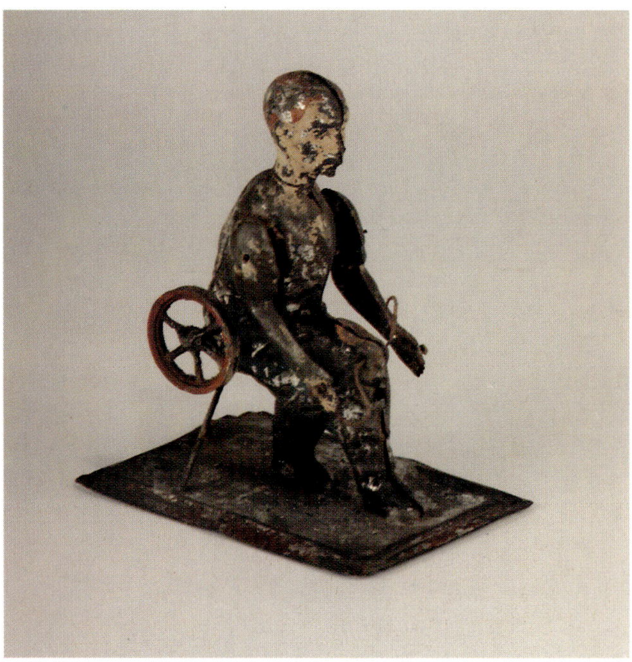

Carette. Shoemaker *(Schuster)*, #684/1, circa 1902. Pulls thread through shoe. Examples of this toy always looks beaten up. 8.8 x 13 x 14.5 cm. $150-250.

Carette. Tailor at sewing machine *(Maschineschneider)*, #699/0, circa 1902. Bing made a similar toy, #9956/197. 9 x 11 x 15.5 cm. $500-750.

Carette. Carousel, #697/5, circa 1905. Three riders on horseback. 14.5d x 21 cm. $500-750. *Provenance Perelman Toy Museum.*

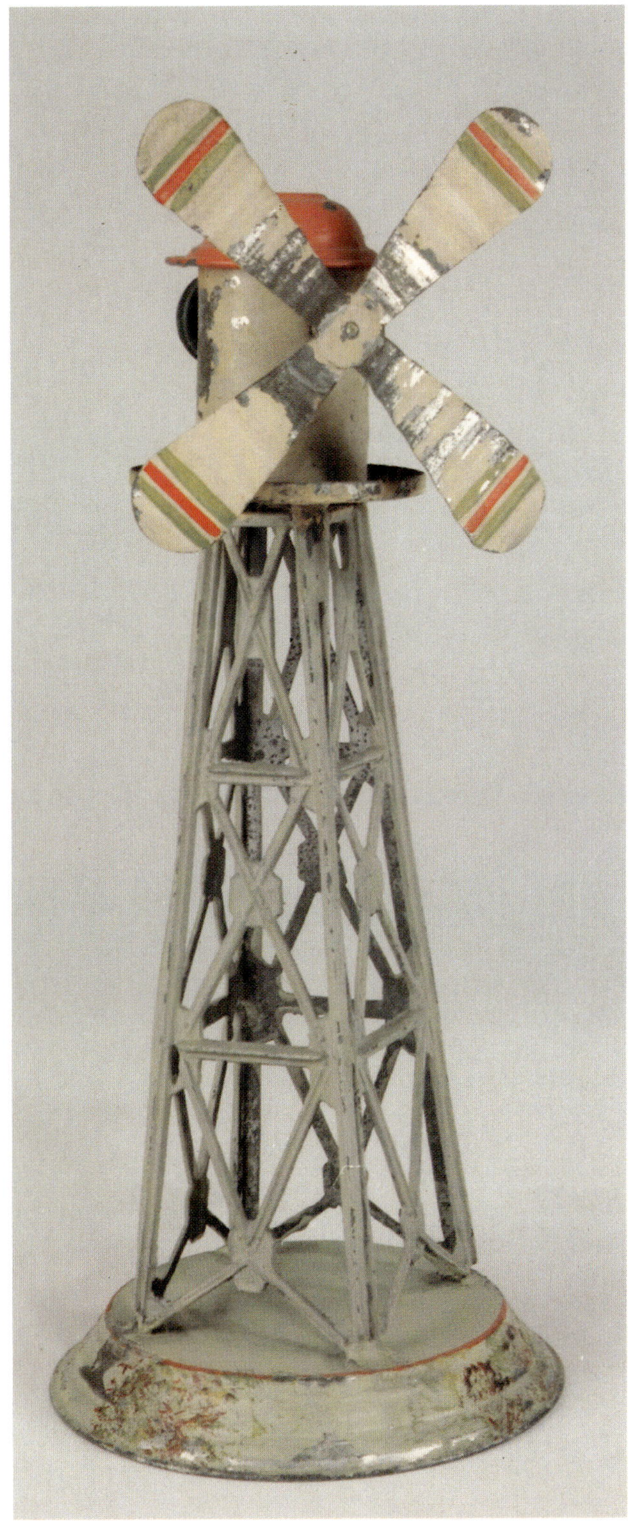

Probably Carette. Windmill, circa 1920. 10d x 26 cm. $50-100.

Probably Carette. Windmill, #697/30A, circa 1902. Also Arnold #161 and Ullmann & Engelmann #2201. 10.2 x 9.8 x 21 cm. $50-100.

Probably Carette. Windmill, circa 1920. Manufacturer could also be Mohr & Krauss. 12.2 x 12.2 x 21 cm. $125-200.

Carette. Windmill, #697/41, circa 1911. 12.5 x 12.5 x 31.8 cm. $50-75.

Probably Carette. Fountain with female figure, circa 1900. 9d x 12.5 cm. $150-250.

Probably Carette. Windmill, circa 1911. 13.5 x 13.5 x 57 cm. Counterweight not visible in picture gives measurement of sail to 20 cm. $200-300.

Probably Carette. Windmill, circa 1911. 12 x 12 x 45.5 cm. $200-300.

Carette. Ballerinas, circa 1890. Girls wear lace tutus and hold hoops in their hands. 14 x 6.4 x 17 cm. $500-750.

Carette. Clown with drum and cymbals *(Paukenschläger)*, #698/31, circa 1911. 12 x 7 x 14 cm. $750-1000.

Carette. Factory *(Workstätte)*, circa 1902. Five figures, five machines, an engine, and a transmission line mounted on a wooden board. 28.6 x 100 x 52.4 cm. $5000-8000. *Courtesy of an anonymous collector.*

Possibly Carette. Boy on swing, circa 1910. Could be Bing or Zschopau School as well. 13.3 x 7.6 x 20 cm. $400-600.

Carette. Troika, circa 1905. Three horses are vertical. 15 x 7.2 x 20 cm. $400-600.

Doll

Nürnberg

Doll & Cie was formed in 1898 by Peter Doll and his partner, J. Sondheim. They manufactured a wide variety of steam engines and accessories. Their toys were of high quality and very sturdy. Doll toys were exported to England, France and the United States. Doll was purchased by Fleischmann in 1938.

Doll. Fountain #601/2, circa 1930. Three tiers, heavy paint loss. 23d x 10 cm. $50-100.

A Doll decal from 1898-1935.

A Doll decal from 1935-1938.

Doll. Fountain *(Fontäne),* #601/4, circa 1930. Shed covers pump. Heavy crazing. 29d x 11 cm. 4.5 cm extra diameter for shed and crank. $50-100.

Doll. Fountain *(Fontäne),* #610/0, circa 1930. 20 x 18 x 6 cm. $50-100.

Doll. Fountain *(Fontäne),* #610/2, circa 1930. Also Moko #5563/3. 28.5 x 28.2 x 8.5 cm. $125-200.

Doll. Hexagonal fountain with four spigots *(Brunnen),* #627/2, circa 1930. 17.5 x 17.5 x 22 cm. $200-400.

Doll. Pump *(Pumpbrunnen),* #902, circa 1930. Has two spigots. 13.6 x 7 x 15.5 cm. $100-150.

Doll. Fountain with two reservoirs *(Doppelbrunnen),* #602/1, circa 1930. 19.5 x 13.5 x 16 cm. $100-200.

Doll. Windmill *(Windmühle),* #633/1, circa 1930. 8 x 8 x 20 cm. $50-100.

Doll. Well *(Ziehbrunnen mit Kippe),* #628/1, circa 1930. Buckets lift water to funnel where it is deposited in the reservoir. 20 x 13 x 23 cm. $75-125.

Doll. Windmill *(Windmühle),* #733/1, circa 1930. 14 x 8.2 x 20 cm. $100-200.

Doll. Windmill *(Windmühle)*, #748/2, circa 1930. Stairs to tower. 11.8 x 11.8 x 27 cm. $150-250.

Doll. Windmill *(Windmühle)*, #748/3, circa 1930. Stairs to tower, guard rail. Open platform at top. Replaced sails. 13.4 x 13.4 x 31.1 cm. $300-500. *Photograph by Françoise R. Hirschberg.*

Doll. Mill with three trip hammers *(Windmühle mit Hammerwerk)*, #707, circa 1930. Replaced sails. 15.8 x 10.8 x 22 cm. $250-350.

Doll. Windmill *(Windmotor mit Pumpe und Schöpwerk)*, #694, circa 1930. Trip hammer, reservoir, and water wheel. Replaced sails. 20.4 x 16.8 x 38 cm. $500-750.

Doll. Windmill with two trip hammers *(Windmotor mit Hammerwerk)*, #668, circa 1930. 13 x 13 x 29 cm. $150-250.

Doll. Windmill *(Windmühle mit Mühlrad und Hammerwerk),* #695. Water wheel, two trip hammers. A large toy. 29 x 19.5 x 46 cm. $1500-2000.

Doll. Dredge *(Bagger)*, #725/0, circa 1930. 17 x 9 x 20 cm. $150-250.

Doll. Vertical sawmill *(Gattersäge)*, #756, circa 1930. 13.7 x 7 x 15.5 cm. $200-300.

Doll. Stamp works, #755, circa 1930. 13.5 x 7 x 11.5 cm. $100-200.

Doll. Punch press *(Exzenterpresse mit Zahnradubersetzung)*, #936, circa 1930. $600-800. *Courtesy of an anonymous collector.*

Doll. Mill *(Mühlhaus mit Hammerwerk),* #788, circa 1930. Water wheel and two trip hammers. 17.5 x 12.2 x 12 cm. $100-200.

Doll. Sawmill with water wheel *(Mühlhaus mit Baumsäge),* #781, circa 1930. 17.5 x 12.2 x 11.5 cm. $150-250.

Doll. Mill *(Mühlhaus mit Pumpwerk und Wasserrad),* #703, circa 1930. The simplest of the Doll mills. 19 x 17.8 x 15 cm. $100-250.

Doll. Mill with fountain *(Mühlhaus mit Fontäne)*, #704, circa 1930. 19.8 x 17 x 14 cm. $250-350.

Doll. Sawmill *(Mühlhaus mit Pumpwerk, Wasserrad und Baumsäge)*, #705, circa 1930. 18 x 17 x 15 cm. $250-350.

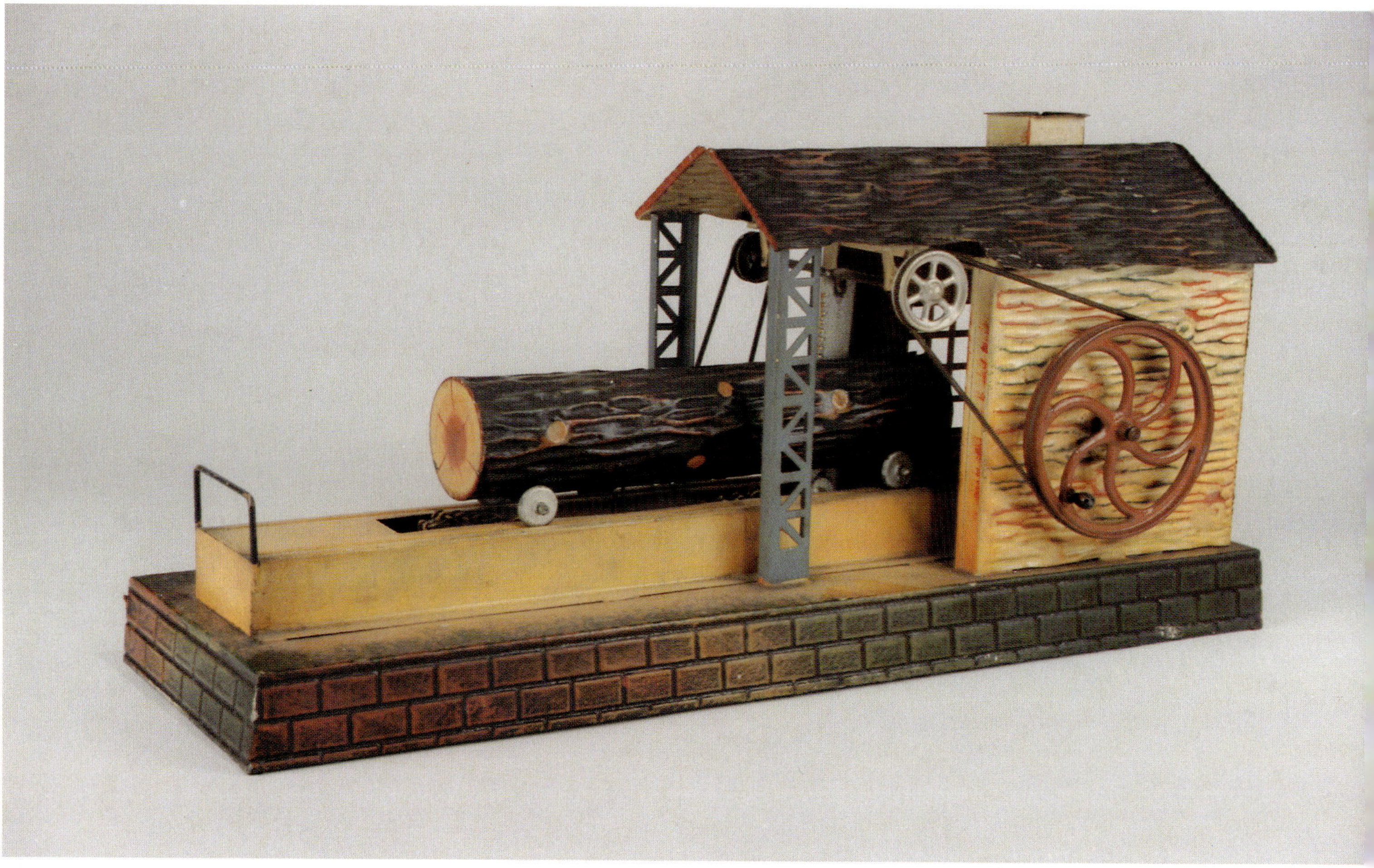

Doll. Sawmill *(Baumsäge)*, #699, ca.1930. Largest of the sawmills. Log rolls in and out of shed. 39.5 x 13 x 20 cm. $500-750.

Doll. Mill *(Mühlhaus mit Pumpwerk, Wasserrad und Hammerwerk),* #706, circa 1930. Water wheel and two trip hammers. 19 x 17 x 15 cm. $150-250.

Doll. Mill with miller and donkey *(Mühlhaus mit Pumpwerk und Wasserrad, Müller und Esel),* #702, circa 1930. Miller and donkey rotate through mill. Water wheel on side. 20 x 20 x 19 cm. $500-750.

Doll. Windmill with miller and donkey *(Windmühle mit Müller und Esel),* #743, circa 1930. Miller and donkey rotate through mill. 16 x 16 x 32 cm. $750-1000.

Doll. Light wheel *(Leuchtrad)*, #974, circa 1930. Made to operate with German battery. Electrified but otherwise original including pink painted base. 17.5 x 12.2 x 22.5 cm. $125-200.

Doll. Rotisserie with chickens *(Hühnerbrateri)*, #916, circa 1930. Surprisingly heavy. 11.5 x 6.6 x 14 cm. $800-1200.

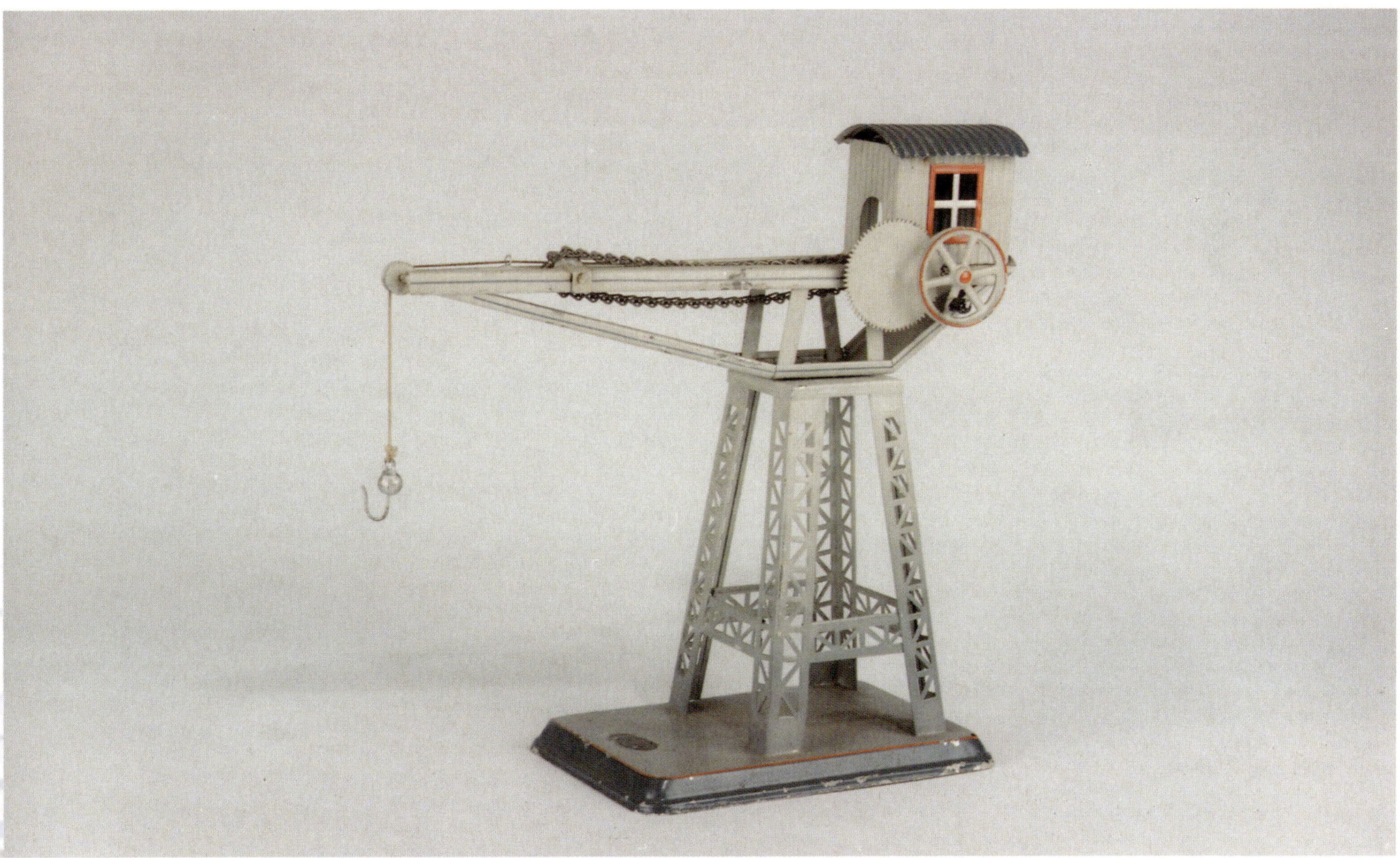

Doll. Crane *(Kran)*, #768, circa 1930. 24.5 x 11.9 x 24.8 cm. Base is 15.9 cm in length. $750-1250.

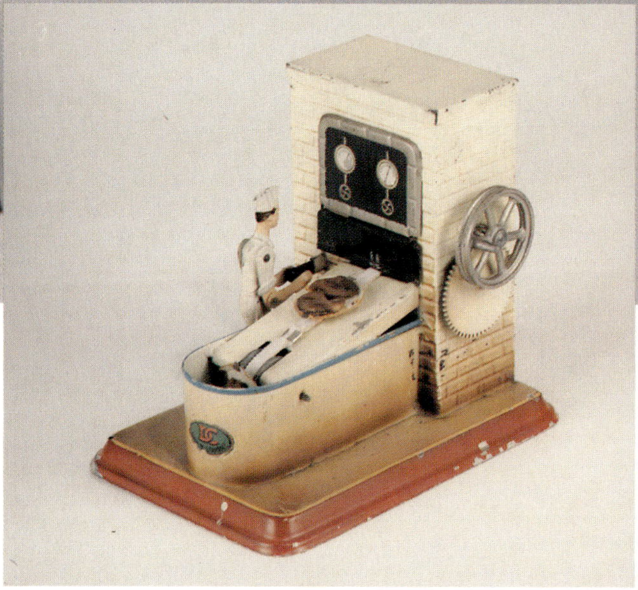

Doll. Baker *(Konditor),* #618, circa 1930. Pasteries and pretzels revolve through oven on belt. 10 x 14 x 13 cm. $300-500. *Provenance Perelman Toy Museum.*

Doll. Men thrashing wheat *(Drescher)*, #619, circa 1930. Flails are articulated. 19 x 12.5 x 15 cm. $1000-1500.

Doll. Ironmonger *(Sensendengler)*, #643, circa 1930. Man hammers thin sheet of metal. 9.4 x 8.5 x 10 cm. $600-800.

Doll. Crane *(Laufkran)*, #766, circa 1930. A large toy. 37 x 12 x 34 cm. $1000-1250.

Doll. Cooper or barrel maker *(Böttcher)*, #644, circa 1930. As the man hammers, he revolves. 14 x 12.2 x 12.5 cm. $350-500.

Doll. Casern *(Kaserne)*, #685, circa 1930. Guards march in and out of sentry post. 18.4 x 10.5 x 13 cm. $1250-1500.

Doll. Relay race (Staffellaufer), #691, circa 1930. Five runners, each holding a baton, run a leg of the race in an Olympic stadium. 22 x 20.1 x 21 cm. $2500-3000.

Doll. Two men working at drop press *(Ramme)*, #690, circa 1930. 17.8 x 15.8 x 20.7 cm. $750-1000.

Doll. Man at pump *(Pumpbrunnen mit Figur)*, #758, circa 1930. Man pumps water into reservoir to his left. 14.8 x 10.2 x 18.5 cm. $800-1200.

Doll. Fire tower with two firemen *(Feuerwehr–Steigerturm)*, #770, circa 1930. Figures, made of lead, disappear into tower and emerge from bottom. 10.1 x 13.2 x 24.5 cm. $800-1000.

Doll. Zoo *(Raubtierzwinger mit Tierwärter und 4 Raubtieren),* #709, ca, 1930. Four Elastolin animals–bear, wolf, tiger, and lion–revolve into animal house while the animal tamer stands guard with his whip. 20 x 19 x 17.5 cm. $1500-2000.

Doll. Logger *(Zimmermann)*, #790, circa 1930. As the man trims the log with his axe, he moves back and forth. 18.6 x 9.5 x 12.5 cm. $1000-1500.

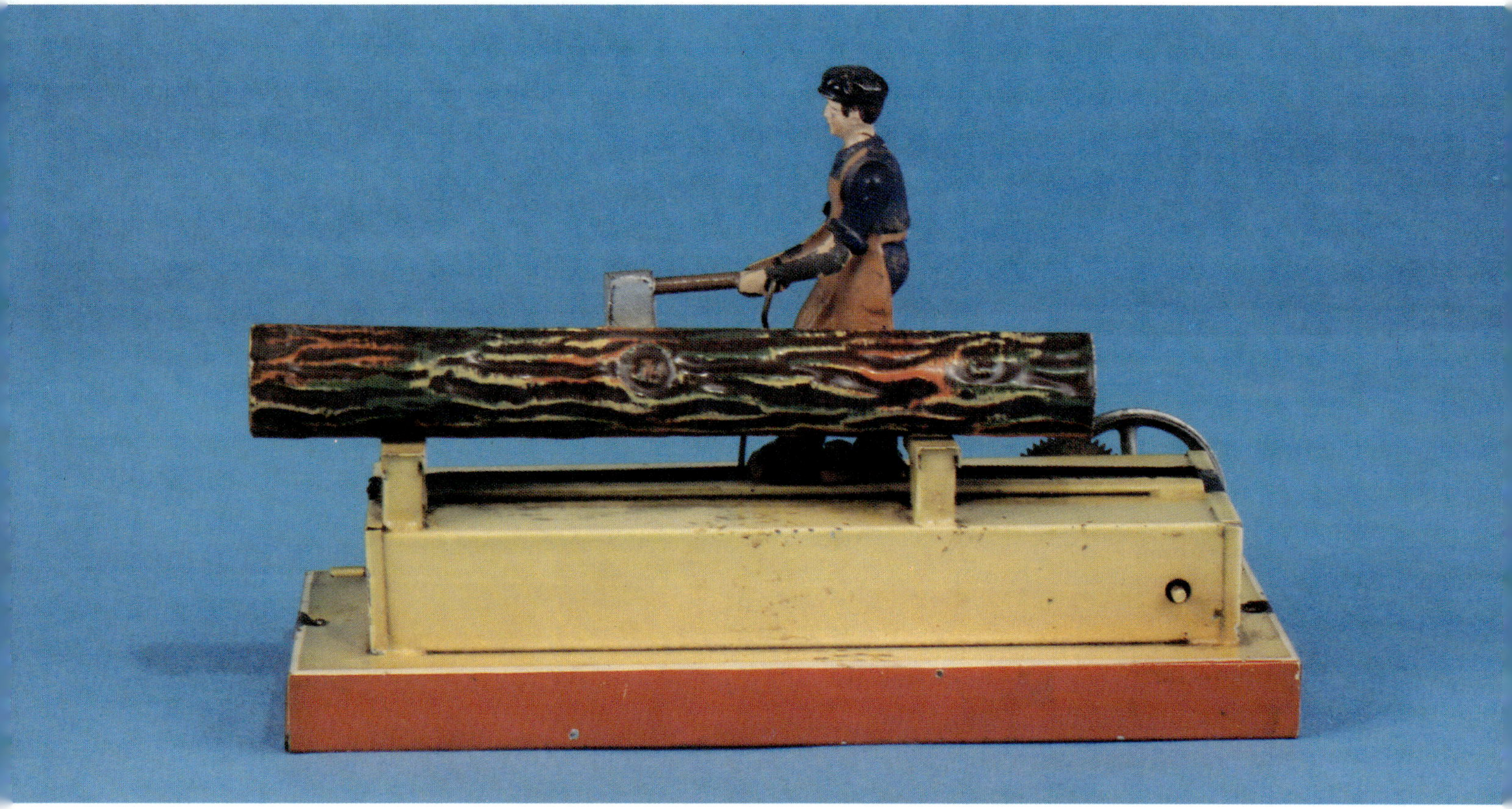

Doll. Logger. Reverse side of toy #790, showing detail.

Doll. Miller *(Müller)*, #791, circa 1930. The miller is lifting a sack of flour from the storage bin. 13.8 x 18 x 17 cm. $1000-1200.

Doll. Miller. Side view of toy #791, showing detail.

Doll. Man at pump *(Mauerpumpe)*, #909, circa 1930. Slight damage to lip of pump. 11 x 8 x 10.5 cm. $50-150.

Doll. Baker at oven *(Bakofen mit Bäcker)*, #922, circa 1930. Electrified with light in rear of toy. 14.3 x 11.2 x 11.7 cm. $500-750.

Doll. Man drinking beer *(Mann am Biertisch)*, #915, circa 1930. 14 x 7.1 x 8.9 cm. $750-1000.

Doll. Street paver *(Pflasterer)*, #925, circa 1930. Man tamps, moving from one side to the other. 13.6 x 14.5 x 12.5 cm. $500-750.

Doll. Man at pump *(Mauerpumpe)*, #926, circa 1930. 8.3 x 15 x 11.2 cm. $800-1000.

Doll. Firemen at pump *(Feuerspritze)*, #927, circa 1930. 17.8 x 7.8 x 9.5 cm. $500-750.

Doll. Chimney sweep *(Schlotfeger),* #962, circa 1930. Sweep enters house and pops up on top. 17.5 x 12 x 21 cm. $500-750. *Provenance Perelman Toy Museum.*

Doll. Chimney sweep *(Schlotfeger),* #961, circa 1930. Sweep disappears into chimney. 14.5 x 10.5 x 21 cm. $600-750.

Doll. Another view of the chimney sweep toy #962, but with sweep in chimney.

Doll. Ship swing *(Schiffschaukel)*, #929, circa 1930. Two boats, each with two figures. 13.6 x 7 x 13.5 cm. $300-500.

Doll. Drummer *(Trommler)*, a variation of #912, circa 1930. Soldier is made of lead. 8.4 x 8.4 x 13.5 cm. $800-1200.

Doll. Soldiers at drill *(Exerzierende Soldaten)*, #687, circa 1930. Soldiers make a half-turn to the left and then to the right. 28 x 12.4 x 12 cm. $1200-1750.

Doll. Sausage maker *(Wurstmaker)*, #910, circa 1930. 13.3 x 9 x 9 cm. $200-300.

Doll. Blacksmith shop with windmill *(Schmiede im Haus mit Windrad)*, #749, circa 1930. 14.5 x 11 x 18 cm. $300-500.

Doll. Blacksmith shop *(Schmiede)*, #720, circa 1930. Shed has electric light. 22 x 12 x 14.5 cm. $350-500.

Doll. Luge *(Rodelbahn),* #789, circa 1930. Two riders. 22.2 x 8 x 16 cm. $1000-1250.

Doll. Luge *(Rodelbahn),* #904, circa 1930. One rider. 13.5 x 7 x 13 cm. $350-500.

Doll. Luge *(Rodelbahn),* #742, circa 1930. Two riders. 33 x 8.6 x 21 cm. $500-750.

Doll. Luge *(Rodelbahn)*, circa 1900. Has also been attributed to Bing. Four riders. Three pines line each side of course. 45.3 x 12.6 x 24 cm. $500-1250.

Doll. Ski lift cars *(Bergbahn)*, #697, circa 1930. Small figures in cars which rotate around station houses. 34.5 x 11.4 x 26.5 cm. $1250-1750.

Doll. Lighthouse *(Leuchtturm)*, #711, circa 1930. Indians with feather headdresses sit in canoes which have sails. Canoes have ogive motion. 27 x 23 x 36.8 cm. $5000-7500. *Photograph by Françoise R. Hirschberg.*

Doll. Flying carousel *(Fliegerkarussell)*, #726, circa 1930. Four swings. Repainted. 17.5 x 12 x 28 cm. $500-750.

Doll. Clown climbing Maypole *(Maibaum)*, #771, circa 1930. 13 x 9.2 x 30.5 cm. $1500-2500.

Doll. Monkey climbing tree to bird house *(Kletteraffe)*, #772, circa 1930. When the monkey reaches the birdhouse, a bird pecks him on the head and he falls to ground. Monkey is a Shucco. 13 x 9 x 30 cm. $1500-2500. *Provenance Perelman Toy Museum.*

Doll. Couple dancing around Maypole *(Maibaum mit Tanzpaar),* #773, circa 1930. Couple is made of celluloid. 15 x 15.5 x 32.3 cm. $2000-3000.

Doll. Flying carousel *(Fliegerkarussell),* #905, circa 1930. Three swings. 9 x 9 x 16 cm. $350-500.

Doll. Flying carousel *(Fliegerkarussell),* #785, circa 1930. Four swings. 12 x 14 x 22 cm. $600-1000.

Doll. Lighthouse *(Leuchtturm),* #712, circa 1930. Modeled after the famous German lighthouse on the Elbe. 19.5 x 19 x 33 cm. $5000-8000. *Courtesy of Louis Natenshon.*

Doll. Elevator *(Personenfahrstuhl),* #645/1, circa 1930. One car with a figure on each side. 12.5 x 10 x 35 cm. $500-750.

Doll. Flying carousel *(Fliegerkarussell)*, #727, circa 1930. Three swings in the shape of airplanes. Tilted canopy. Stairway to swings. 12 x 17.5 x 28 cm. $2000-3000.

Doll. Ferris wheel *(Russisches Karussell),* #908, circa 1930. Three swings. 8 x 8 x 19.5 cm. $300-500. *Photograph by Françoise R. Hirschberg.*

Doll. Ferris wheel *(Russisches Karussell),* #729/1, circa 1930. Four seats. 12.8 x 12.8 x 28 cm. $350-700.

Doll. Ferris wheel *(Russisches Karussell),* #787, circa 1930. Five seats. 9.2 x 13 x 22 cm. $750-1000.

Doll. Ferris wheel, circa 1930. Six seats. Repainted. Replaced figures. 15.7 x 15.7 x 31 cm. $750-1500.

Doll. Ferris wheel *(Russisches Karussell)*, #729/2, circa 1930. Six seats, each holding two riders. Old repaint, but original figures and flag. 17.5 x 17.5 x 35.2 cm. $2000-2500.

Doll. Ferris wheel *(Russisches Karussell),* #729/3, circa 1930. Six seats, each holding two riders. All original. Superb. Also comes in a variation with scalloped seats. See King's book for an example from the British-Museum. 24 x 18.5 x 38 cm. $4500-6000.

Doll. Ferris wheel *(Russisches Karussell),* #729/5, circa 1930. Six seats, each holding four riders. Detachable stairs lead to waiting platform. The largest of the Ferris wheels. 29.5 x 36.6 x 53.3 cm. $8000-12000.

Doll. Reverse side of toy #729/5, showing view of platform and detachable stairs.

Doll. The whip, circa 1930. Carnival ride with six cars, each holding two riders. As platform rotates, the cars move in and out. A rare toy. 33 x 31.5 x 8 cm. $4000-6000.

Eberl

Nürnberg

Hans Eberl founded this novelty toy company, which existed from 1902 until 1929. I know of only one steam toy which they manufactured.

Eberl. Reverse side of the elephant toy, showing detail.

Eberl. Indian elephant with rider and music box, circa 1906. Elephants head moves up and down and trunk appears to wind music box. 22.5 x 8.7 x 16.8 cm. $1500-2000.

Falk

Nürnberg

Joseph Falk, once an employee of Carette, formed his own business around 1895. He was a large producer of steam toys, many of which are wrongly attributed to Bing or to Doll. There is a great similarity among the mills and Ferris wheels which causes the confusion, but both dimension and color variations lead to proper identification. Falk took over Schoenner around 1910 and was himself absorbed by Plank in the early 1930s.

Mark used by Falk, 1900-1935.

Falk. Windmill *(Windmühle),* #214/0, circa 1930. 7.2d x 15.5 cm. $50-100.

Falk. Ferris wheel *(Russische Karussell),* #218/1, circa 1930. Two seats. 10 x 8 x 23 cm. $400-600. *Photograph by Françoise R. Hirschberg.*

Falk. Ferris wheel *(Russische Karussell),* #218/2, circa 1930. Four seats with original composition figures and flag. 15.6 x 15.6 x 26 cm. $600-800. *Photograph by Françoise R. Hirschberg.*

Falk. Ferris wheel *(Russische Karussell),* #218/3, circa 1930. Six seats with original composition figures and flag. 20 x 16.5 x 29.8 cm. $1000-1500.

Falk. Fountain, #224, circa 1930. Three tiers, paint loss. 28 x 24.2 x 14 cm. $250-350.

Falk. Windmill *(Windmühle)*, #223/1, circa 1930. 12.2 x 12.2 x 27 cm. $200-300.

Falk. Dredge *(Baggerwerk)*, #225/2, circa 1930. Among the finest of the Falk dredges. 34 x 20 x 28.3 cm. $700-1000. *Photograph by Françoise R. Hirschberg.*

Falk. Water elevator *(Schöpfwerk)*, #237/1, circa 1930. 21.6 x 19 x 14 cm. $350-500.

Falk. Boat swings *(Schiffschaukel)*, #246/2, ca 1925. Two swings, replaced flags. 18 x 9.4 x 20 cm. $300-500.

Falk. Sawmill *(Baumsäge)*, #248, circa 1930. 14.5 x 9 x 5.5 cm. $50-100.

Falk. Sawmill *(Wassermühle mit Baumsäge)*, #253, circa 1930. Mill with water wheel. Log being cut by saw. 26 x 16.4 x 16.5 cm. $350-500.

Falk. House with windmill and pump *(Haus mit Windmotor und Brunnen)*, #260, circa 1930. 24 x 16.4 x 27.5 cm. $500-750.

Falk. Windmill with saw and water wheel, #275, circa 1930. 16.5 x 18.5 x 23.5 cm. $250-400.

Falk. Double swing (Doppelschaukel), #238, circa 1930. The action is more like a teeter-totter than a swing. As one seat with rider rises, the other lowers. 15 x 8.2 x 12.5 cm. $300-500.

Falk. Fountain, #284/3, circa 1930. Missing one rail. 23.5 x 23.5 x 8.2 cm. $100-200.

Falk. House with windmill, water wheel and trip hammer, #286/67 or 286/88, circa 1930. 17.2 x 11.8 x 26 cm. $125-200.

Falk. Tinsmith shop *(Klempnerwerkstatt)*, #286/63, circa 1930. Very similar to the earlier Mohr & Krauss version. 14 x 8.2 x 12.5 cm. $400-600.

Falk. Ferris wheel, #286/78, circa 1930. Simple Ferris wheel with six orange and green seats. 12 x 9.6 x 21 cm. $75-150.

Falk. Man with circular saw and work table, #299/13, circa 1930. Note the strong resemblance to the slightly earlier toys by Arnold. 17.6 x 9 x 11.5 cm. $200-250.

Falk. Ferris wheel with centrimotor, circa 1930. Simple Ferris wheel with six yellow and green seats. 17.8 x 9.7 x 21 cm. $150-250.

Probably Falk. Dredge *(Kettenbagger)*, circa 1925. Three buckets in well. Similar to the #217 series but with a curved roof. 9.5 x 20.5 x 19 cm. $150-200. *Photograph by Françoise R. Hirschberg.*

Falk. Sawyers *(Handwerker)*, #299/14, circa 1930. Man at left is using a bow saw while the man at the right is using a circular saw. Note the strong resemblance to the slightly earlier toys by Arnold. 24 x 11 x 11.5 cm. $200-250.

Falk. Pump with two reservoirs *(Brunnen),* circa 1930. 16.2 x 10 x 21 cm. $150-250.

Falk. Printing press, circa 1900. Like many of the earlier toys, this toy is mounted on a wooden base, with maker's mark attached separately. 16 x 10.4 x 12.5 cm. $400-600.

Probably Falk. Marble mill with well, circa 1930. Marbles are brought to top ramp by ledges attached to chain. Paint loss at base. 29.2 x 10.2 x 26.5 cm. $300-500.

Fleischmann

Nürnberg

Founded in Nuremburg by Jean Fleischmann in 1877, the production of steam toys began largely after 1945. However, Fleischmann absorbed the firm of Georg Staudt in 1928 and took over many of the doll models in 1938.

Fleischmann. Windmill *(Windmühle)*, #210, circa 1950. 8.5 x 8.5 x 20 cm. $50-100.

Fleischmann. Dredge with three buckets *(Schöpfwerk)*, #225, circa 1950. 17.6 x 10.2 x 15.8 cm. $75-150.

Fleischmann. Man turning water wheel *(Schöpfrad)*, #226, circa 1950, 14 x 10.5 x 15 cm. $100-200.

Fleischmann. Mill with trip hammers *(Mühlhaus mit Hammerwerk)*, #230, circa 1950. 18.2 x 15.4 x 12.5 cm. $75-150.

Fleischmann. Blacksmith at anvil *(Schmied am Amboss),* #232, circa 1950,. 13.9 x 7.4 x 10 cm. $75-150.

Fleischmann. Man at grindstone *(Schleifstein),* #235, circa 1950. 8.4 x 8.4 x 10 cm. $25-50.

Fleischmann. Blacksmith at anvil in shed *(Schmiede-Werkstatt),* #233, circa 1950. 14 x 11 x 11 cm. $125-200.

Fleischmann. Sausage-maker *(Wurstmaschine),* #236, circa 1950. 13.3 x 8.9 x 10 cm. $75-200.

Fleischmann. Man at well *(Schöpfbrunnen),* #237, circa 1950. 13.3 x 9.3 x 16 cm. Well has two buckets. $100-200.

Fleischmann. Circular saw *(Kreissage)*, #249, circa 1950. 13.2 x 10.2 x 9 cm. $50-100.

Fleischmann. Castle with two swans swimming in moat, circa 1925. 10d x 16.5 cm. $500-750.

Folk Art

Folk art. Homemade foresters with axes at a workbench, circa 1925. Man on left smoking a cigar, man on right smoking a pipe. Painted wood. 26 x 9.2 x 21.5 cm. $200-300.

Folk art. Homemade, circa 1925. Woman at left with axe. Man at right with bow saw. Painted wood. 25 x 8.5 x 22 cm. $200-300.

Folk art. Homemade tailor shop, circa 1925. Tailor at left sits cross-legged sewing. Tailor on right pressing material. Painted wood. 22 x 9 x 17.5 cm. $200-300.

Günthermann

Nürnberg

Siegfried Günthermann founded this Nürnberg firm in 1877. After his death in 1890, the company was run by his widow and her second husband and Günthermann's two sons. The company's toys were charming, especially their carousels. Their steam toys followed their wind-up counterparts. Much of the Günthermann production was made for export to the United States. The firm operated until 1965.

Günthermann. Woman sewing, circa 1900. This toy is usually seen as a wind-up. 13.8 x 10.2 x 14 cm. $1000-1500. *Photograph by Françoise R. Hirschberg.*

Günthermann. Woman at spinning wheel, circa 1900. 17.7 x 8.9 x 18 cm. $1200-1500.

Günthermann. Three-horse carousel, circa 1910. Horses and riders. Repainted with replaced flag. 19d x 41.2 cm. $600-800.

Günthermann. Four-horse carousel, circa 1910. Horses and riders. 17d x 35.2 cm. $600-800.

Probably Günthermann. Carousel with horses and swings, circa 1910. Three horses with riders in body of the toy and 6 swings on the canopy. 21d x 29.5 cm. $600-800.

Probably Günthermann. Four-horse carousel, circa 1910. Horses and riders. The Washington Doll Museum has one just like this, down to the mustard color paint on the platform. 18d x 28cm. $600-800.

Hawkotte and Klausman

United States

Not much is known about H&K, which operated as a novelty company in either Indiana or Ohio during the 1920s. Their toys are bulky and I must confess far from attractive.

Hawkotte & Klausman. Merry-go-round with six seats, circa 1920. 23.2d x 38 cm. Mounted on a plastic board with an electric motor. The seats are red, white and blue. This toy did not come with figures. $400-600.

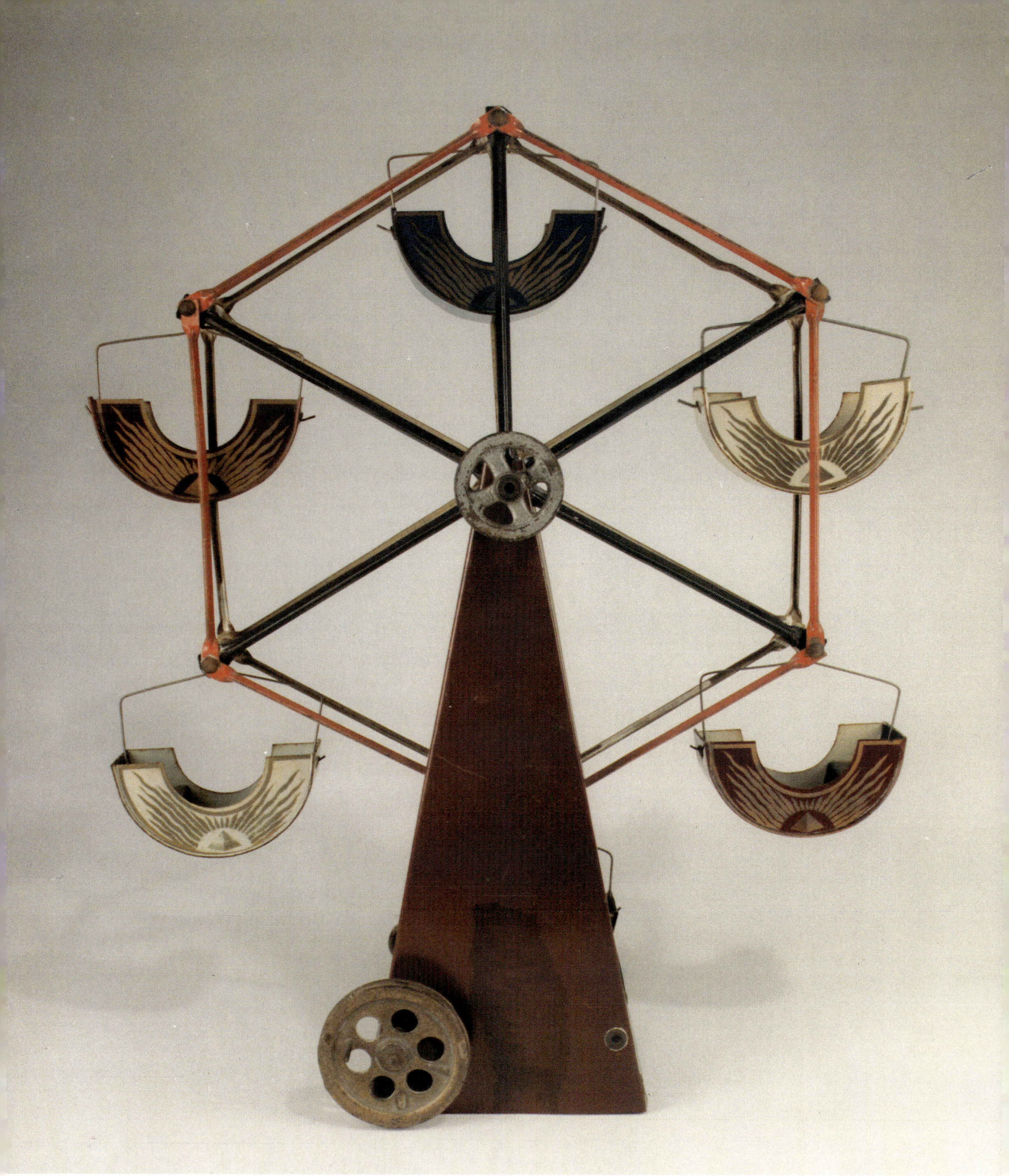

Hawkotte & Klausman. Ferris wheel with six seats, circa 1920. 47.5 x 17.5 x 53.3 cm. Seats are red, white and blue. This toy did not come with figures. $500-750.

Ives

United States

Ives is one of the great manufacturers of clockwork and toy trains. Fortunately for us, he made a number of marvelous steam and hot air toys which are paper lithography or lithography on wood. Edward Riley Ives began his efforts in Plymouth, Massachusetts. Later he moved to Bridgeport, Connecticut, where he produced toys under the name of Ives, Blakeslee, and Williams. Ives companies operated from 1866 to 1910 as general toy producers, and after that as train manufacturers, until they were purchased by Lionel. Their slogan was "Ives Toys Makes Happy Boys."

Ives. Butcher shop, #100-1, circa 1900. Paper lithography on wooden base. 17 x 4.4 x 16.5 cm. $600-800.

Ives. Sawyer with bow saw, #72, circa 1900. Paper lithography. Hot air toy on a wooden base. Replaced blue fan. 16.4 x 3.4 x 19 cm. $400-600.

Ives. Sawyer with bow saw in shed, #100-2, circa 1869. Paper lithography on wooden base. 17 x 2.5 x 16 cm. $300-500. *Photograph by Françoise R. Hirschberg.*

Ives. Washerwoman with tub in shed, #100-4, circa 1878. Paper lithography on wooden base. 17 x 2.8 x 16.7 cm. $400-600.

Ives. Sawyer with bow saw, circa 1900. Paper lithography on wood. Wooden base. 24 x 7.5 x 24 cm. $400-600.

Ives. Blacksmiths at anvil, circa 1900. Paper lithography on wood. Wooden base. 23 x 9 x 16 cm. $400-600.

Ives. Puss in Boots, #150, circa 1900. Paper lithography on wooden base. Cat's eyes move. 17 x 2.2 x 13.2 cm. $500-750.

Ives. Friar ringing church bell, circa 1900. Paper lithography on wood. Wooden base. 25 x 6 x 27 cm. $400-600.

Ives. Shoemaker, #154, circa 1900. Paper lithography on wooden base. 13 x 2.2 x 13 cm. $400-600.

Kohnstam

Fürth

Moses Kohnstam was a jobber and retailer of toys. Although the company itself produced no toys, it carried an extensive line of toys. Kohnstam gave his own catalog numbers to the toys that appeared in his catalogs. Where possible these will be cited. Kohnstam operated from 1875 until 1933. The company is also known as "Moko."

Krauss

Nürnberg

A Krauss mark, 1923-1938.

There were a succession of companies which operated as Krauss, Mohr & Company from 1895 to 1903, Mohr and Krauss from 1901 until 1923, and Wilhelm Krauss from 1923 until 1938. The various companies made simple-style, colorful toys. They produced merry-go-rounds (roundabouts) as well as the much-sought slate dancers. Krauss toys are found in Ullmann and Engelmann catalogs using the latter's catalog numbers. Many inexpensive Krauss toys were made for export to the United States.

A Krauss mark, 1933-1938.

Mohr & Krauss. Windmill with centrimotor, circa 1905. Walking beam motion. The sails have a Falk look to them. 25 x 9 x 25 cm. $100-150.

Mohr & Krauss. Windmill and water works, circa 1905. Walking beam motion. Notice the bird on the roof of the shed. 32.4 x 11.2 x 30.5 cm. $400-600.

Mohr & Krauss. Group of three drop hammers, circa 1920. (At left) Trip hammers. 11.5 x 8.7 x 13 cm. $25-75. (At center) Drop press. 6.7 x 6.7 x 11 cm. $25-75. (At right) Drop hammer. 9 x 8.5 x 13 cm. $25-75.

Mohr & Krauss. Factory, circa 1905. Vertical saw on the left, grinder and circular saw in the center, and drop hammers on the right. Repainted. 40.6 x 15 x 28.5 cm. $100-250.

Mohr & Krauss. Windmill, circa 1920. 7.6 x 7.6 x 22 cm. $25-50. *Photograph by Françoise R. Hirschberg.*

Mohr & Krauss. Windmill, circa 1920. Simple but with charm. Comes in several color variations; sometimes the window decal is on the other side. 8.2 x 8.2 x 19.5 cm. $50-100.

Wilhelm Krauss. Group of three windmills, circa 1946. (At left) 9 x 6.6 x 12 cm. $20-50. (At center) 11.2 x 8.5 x 17 cm. $50-75. (At right) 11.2 x 8.5 x 20 cm. $50-75.

Probably Wilhelm Krauss. Windmill, circa 1930. Very large. 16 x 22.5 x 36.8 cm. $75-150.

Mohr & Krauss. Windmill, circa 1905. The sails have a Falk look to them. Hexagonal shape. 10 x 10 x 18 cm. $50-100.

Probably Mohr & Krauss. Windmill, circa 1905. Ornate circular railing. 12.5d x 21 cm. $50-100.

Mohr & Krauss. Windmill and well, circa 1905. The bucket in the well is raised and lowered by the large wheel under the windmill. Repainted. 23.5 x 12.5 x 31.8 cm. $150-250.

Krauss, Mohr & Co. Windmill with reservoirs and pumps, circa 1900. Could easily be a Carette. Repainted. 26 x 11 x 34.9 cm. $200-300.

Mohr & Krauss. Windmill and well, circa 1905. Buckets attached to wheel. 21.3 x 9 x 17 cm. $100-200.

Mohr & Krauss. Windmill and well, circa 1905. Buckets are on chain. 25 x 9.4 x 17.5 cm. $100-200.

Mohr & Krauss. Factory works, circa 1912. Mill sits in front of mountain. These inexpensively made toys were not primed; the result is that they often flake. 22 x 17.2 x 15 cm. $75-150. *Photograph by Françoise R. Hirschberg.*

Mohr & Krauss. Refinery, circa 1912. Mill sits in front of mountain. Notice slag heap in reservoir. These inexpensively made toys were not primed. 22 x 17.5 x 15 cm. $150-250.

Mohr & Krauss. Water works, circa 1905. Mill sits in front of mountain. Notice small cottage on left side. Water spout is almost top middle. 15.7 x 15.4 x 17.5 cm. $125-200.

Mohr & Krauss. Fountain, circa 1905. A lion's head juts out of the mountain. 13.5 x 13.2 x 15.5 cm. $50-100.

Mohr & Krauss. Dredge and windmill *(Baggerwerk)*, circa 1905. Similar to Bing #9956/21, circa 1908. 23.6 x 10.6 x 20 cm. $125-250.

Mohr & Krauss. Water works, circa 1905. This is a much larger version of the preceding toy. Mill sits in front of mountain. Notice small cottage on left side. 22.3 x 22.3 x 22.5 cm. $500-750.

Mohr & Krauss. Mill with trip hammer, circa 1905. 11.8 x 13.6 x 12 cm. $50-75.

Mohr & Krauss. Mill with trip hammers. Notice difference in water wheel on left and presence of pear tree. 17.4 x 11.4 x 12 cm. $100-200. *Photograph by Françoise R. Hirschberg.*

Mohr & Krauss. Mill with trip hammers. Also Falk #286/69, Moko #5561, and Hausmann #949/6. 18 x 12 x 12 cm. $50-100.

Mohr & Krauss. Mill with trip hammers, circa 1920. Mountain in back provides water power. 19 x 14 x 15.5 cm. $50-100.

Mohr & Krauss. Stamp works, circa 1905. Three trip hammers and nice lithography. 17 x 7.8 x 15 cm. $150-250.

Mohr & Krauss. Stamp works, circa 1905. Similar to preceding toy but with large pear tree. 25.5 x 11.5 x 15 cm. $100-250.

Mohr & Krauss. Stamp works. Three trip hammers, pear tree and pair of doves sitting on roof. 21 x 11 x 15.5 cm. $250-500.

Mohr & Krauss. Saw mill, circa 1905. A somewhat rare saw mill. 26.5 x 14.4 x 15 cm. $350-500.

Probably Wilhelm Krauss. Saw mill, circa 1930. A somewhat rare saw mill. 29 x 16 x 17 cm. $350-500.

Mohr & Krauss. Workman cooking, circa 1905. Man with frying pan standing in front of cottage, mountain backdrop. A rare and interesting piece. Some repainting on cottage. 19.6 x 14.8 x 13.5 cm. $100-200. *Photograph by Françoise R. Hirschberg.*

Mohr & Krauss. Water mill with trip hammer, circa 1920. Some repainting. 17.2 x 11.4 x 11.5 cm. $50-75.

Mohr & Krauss. Water mill with trip hammer, circa 1920. Also Falk #286/66, circa 1925. 17.6 x 11.8 x 11.5 cm. $50-100.

Wilhelm Krauss. Water mill, circa 1935. Centrimotor at left, pear tree on right. 24 x 11.5 x 12.5 cm. $125-200.

Wilhelm Krauss. Water mill, circa 1935. Stamp works to left in front of tree, and trip hammer. 24 x 12.9 x 18 cm. $125-200.

Mohr & Krauss. Sawyer, circa 1905. Trip hammer and pear tree at left, windmill in center and sawyer with bow saw at right. 34.3 x 11.4 x 24 cm. $400-600.

Mohr & Krauss. Windmill, circa 1905. Pump and reservoir at left, donkey in front of pear tree in center, and mill at right. 34.3 x 11.5 x 23 cm. $400-600.

Mohr & Krauss. Sawyer, circa 1905. Sawyer has bow saw and is driven by centrimotor. 21.2 x 8 x 13 cm. $125-200.

Wilhelm Krauss. Sawyer with bow saw, circa 1935. 11.3 x 7.8 x 11.5 cm. $50-100.

Mohr & Krauss. Waterworks with windmill, circa 1920. Centrimotor at left in front of pear tree, water mill with bird on roof, and walking beam windmill. 33 x 11.8 x 24.5 cm. $400-600. *Photograph by Françoise R. Hirschberg.*

Wilhelm Krauss. Grinder, circa 1935. 11.3 x 7.8 x 11 cm. $50-100.

Wilhelm Krauss. Man at stamping press, circa 1935. 11.3 x 7.8 x 11 cm. $50-100.

Wilhelm Krauss. Sawyers, circa 1930. These are probably harder to come by than the Bing counterparts. 21 x 8.2 x 12.5 cm. $250-400.

Wilhelm Krauss. Sawyer with bow saw, circa 1930. 13.8 x 8 x 13 cm. $100-200.

Wilhelm Krauss. Man at drill press, circa 1930. 14 x 8.2 x 14.5 cm. $100-200.

Wilhelm Krauss. Two color wheels, circa 1930. Left, 11.5 x 7 x 13 cm. $200-300. Right, 14.5 x 8.5 x 15.5 cm. $200-300.

Wilhelm Krauss. Swing seats, circa 1930. Four boats each with a rider. 24 x 10 x 14.5 cm. $500-750.

Wilhelm Krauss. Happy Jack and Happy James slate dancers, circa 1925. Notice small stage with their names behind them. This is the only steam toy which has names on the stage. 13 x 7 x 16.5 cm. $1000-1500.

Wilhelm Krauss. Harlequin slate dancer, circa 1930. Notice small stage, characteristic of the slate dancers. 9 x 7 x 14 cm. $300-500.

Wilhelm Krauss. Two comedians slate dancers, circa 1925. Pat and Patachon. The comedians may also have the Scandinavian names Fyrtaarnet and Bivøgnen. Notice small stage behind them. 18.2 x 7 x 20 cm. $500-750. *Provenance Gottschalk Collection.*

Wilhelm Krauss. Carousel, circa 1925. Also Moko. Flat-figured three-horse carousel. 12 x 9.5 x 15 cm. $200-400.

Wilhelm Krauss. Charlie Chaplin slate dancer with centrimotor, circa 1925. He also comes in a version without the centrimotor on a smaller base. Notice small stage behind him. 17 x 9 x 17 cm. $1500-2500.

Perhaps Wilhelm Krauss. Mickey Mouse slate dancer, circa 1930. Could be by Distler. The royalty of slate dancers. One of only a half dozen known. $15000 and up. *Photograph by Michael Floman. Courtesy of Carl Lobel.*

Wilhelm Krauss. Carousel, circa 1925. Also Moko #5348. Four brightly colored gondolas. 17 x 11.6 x 24 cm. $400-600.

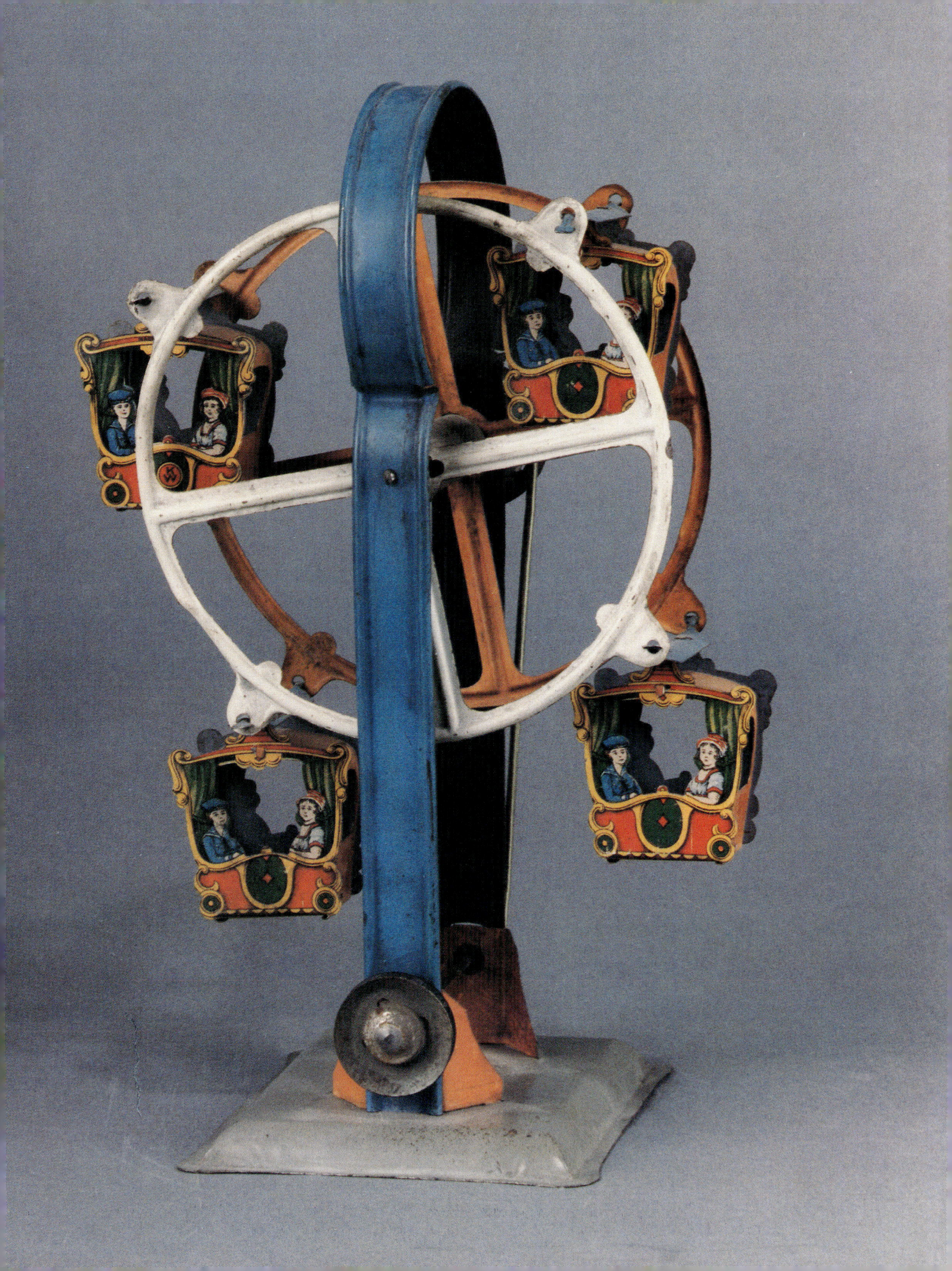

Wilhelm Krauss. Carousel *(Russisches Carussell),* circa 1925. A more elaborate version of the previous toy, with entryway. Four brightly colored gondolas. 18 x 12.4 x 24 cm. $1000-1500.

Mohr & Krauss. Shoemaker's apprentice, circa 1910. 14 x 8.2 x 12.5 cm. $400-600.

Krauss, Mohr & Co. Strongman, circa 1895. Hercules with barbell. 10 x 7.6 x 9.5 cm. $300-500.

Mohr & Krauss. Shoemaker, circa 1910. 14 x 8.2 x 12.5 cm. $400-600. *Provenance Gottschalk Collection.*

Krauss, Mohr & Co. Acrobats, circa 1900. Two men on high bar, each facing different direction. 14.3 x 8 x 16 cm. $250-400.

Mohr & Krauss. Tinsmith with centrimotor, circa 1910. 14 x 8.2 x 12.5 cm. $400-600.

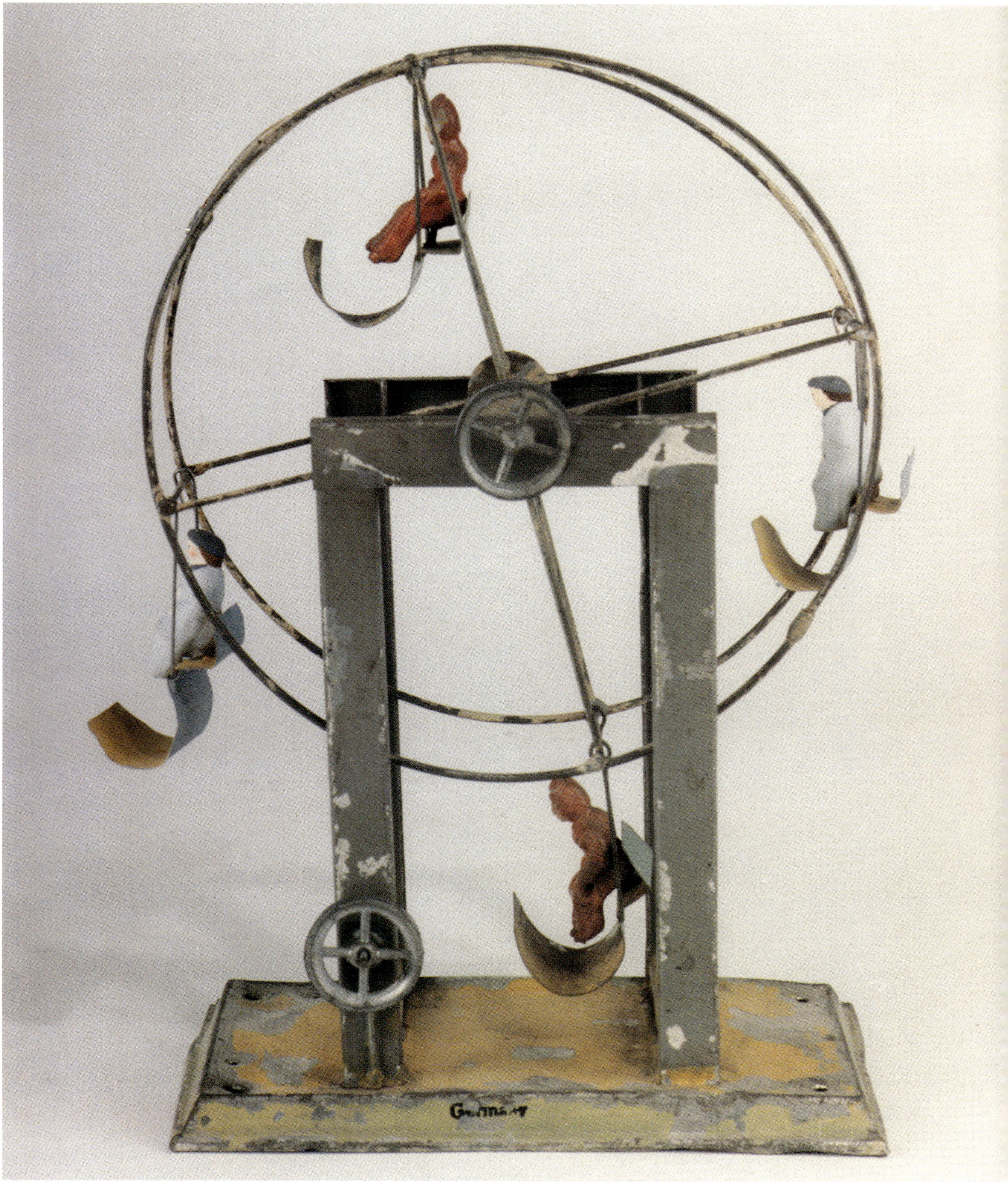

Krauss, Mohr & Co. Carousel, circa 1900. A simple four-seat carousel. Two original red figures. 15.5 x 9 x 23.5 cm. $100-250.

Krauss, Mohr & Co. Ferris wheel, circa 1900. An elaborate toy with four gondolas which spin as the toy rotates. Some repainting. 22.5 x 22.1 x 42.8 cm. $1200-1500.

Mohr & Krauss. Ice skating rink *(Eislaufbahn),* circa 1925. Five lead figures rotate around the rink. 29.5 x 21.4 x 17 cm. $2000-3000.

Wilhelm Krauss. Flying carousel, circa 1930. Three yellow dirigibles. Also made as a wind-up. 12 x 12 x 18 cm. $1000-1500. *Photograph by Françoise R. Hirschberg.*

Wilhelm Krauss. Flying carousel, circa 1930. Three Folker air-
planes. 11.6d x 24.5 cm. $600-800.

Wilhelm Krauss. Flying carousel, circa 1930. Three sailors in swings. Also made as a wind-up. 11.8d x 25.5 cm. $600-800.

Wilhelm Krauss. Flying carousel, circa 1930. Four Distler or Maier penny toy dirigibles. Also made as a wind-up. Superb. 11 x 11 x 25 cm. $1500-2000.

Märklin

Göppingen

Märklin is undoubtedly the premier manufacturer of steam toys. Regardless of the value of other toys made by other manufacturers, Märklin continues to get stronger. Founded in 1859 by Theodor Märklin, the company is still in business manufacturing small-gauge toy trains. Märklin produced a wide variety of toys, the prize examples of which (the Ferris wheel and circus carousel) are shown in this volume.

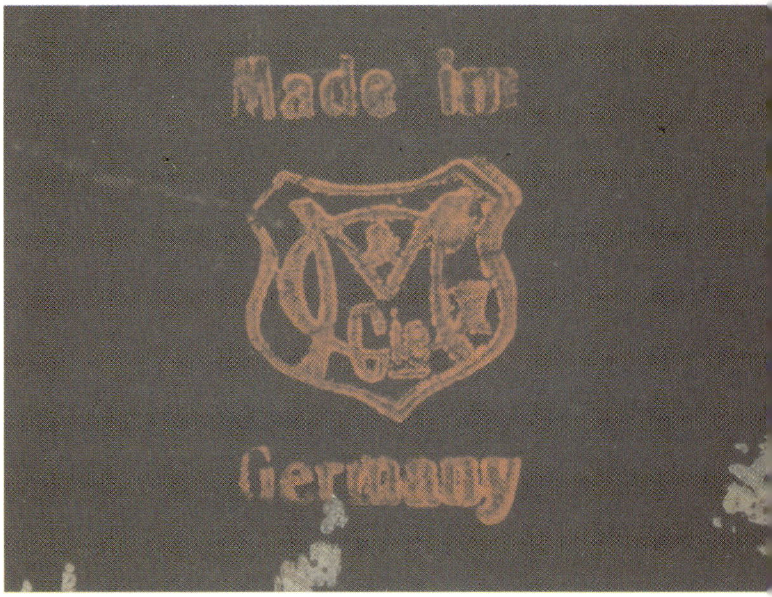

A Märklin mark used circa 1929.

A Märklin mark used after 1930.

Märklin. Sorter *(Getreide-Sortiermaschine oder Trieur)*, #4207, circa 1902. Replacement bucket. The toy originally came with two buckets and two woven bushel baskets. 23 x 9.6 x 16 cm. $750-1000.

Märklin. Thrashing machine *(Dreschmaschine),* #4201, circa 1902. A superb example of the rare thrashing machine on wheels. Only known example to say 'Thrashing-machine" in English. 29 x 10 x 22 cm. $7500-12500. *Courtesy of Brent Rowell.*

Märklin. Reverse side of thrashing machine. *Courtesy of Brent Rowell.*

Märklin. Fountain *(Fontäne)*, #4224, circa 1904. Fountain with six porpoises on edge. Pedestal has four porpoises on top and lion heads on sides. Missing female figure in center. One of the two or three finest fountains made by Märklin. 38 x 32 x 19 cm. $2500-5000.

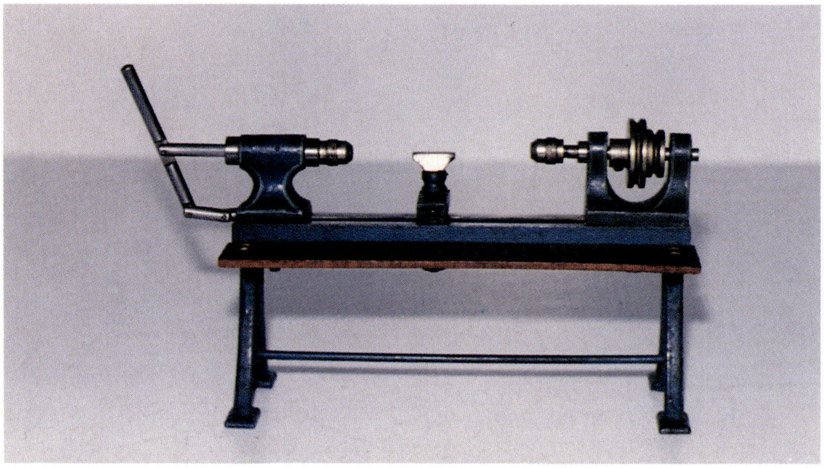

Märklin. Lathe *(Drehbank)*, #4262, circa 1909. 25.4 x 7.9 x 13.3 cm. $1200-1500. *Courtesy of an anonymous collector. Photograph by Françoise R. Hirschberg.*

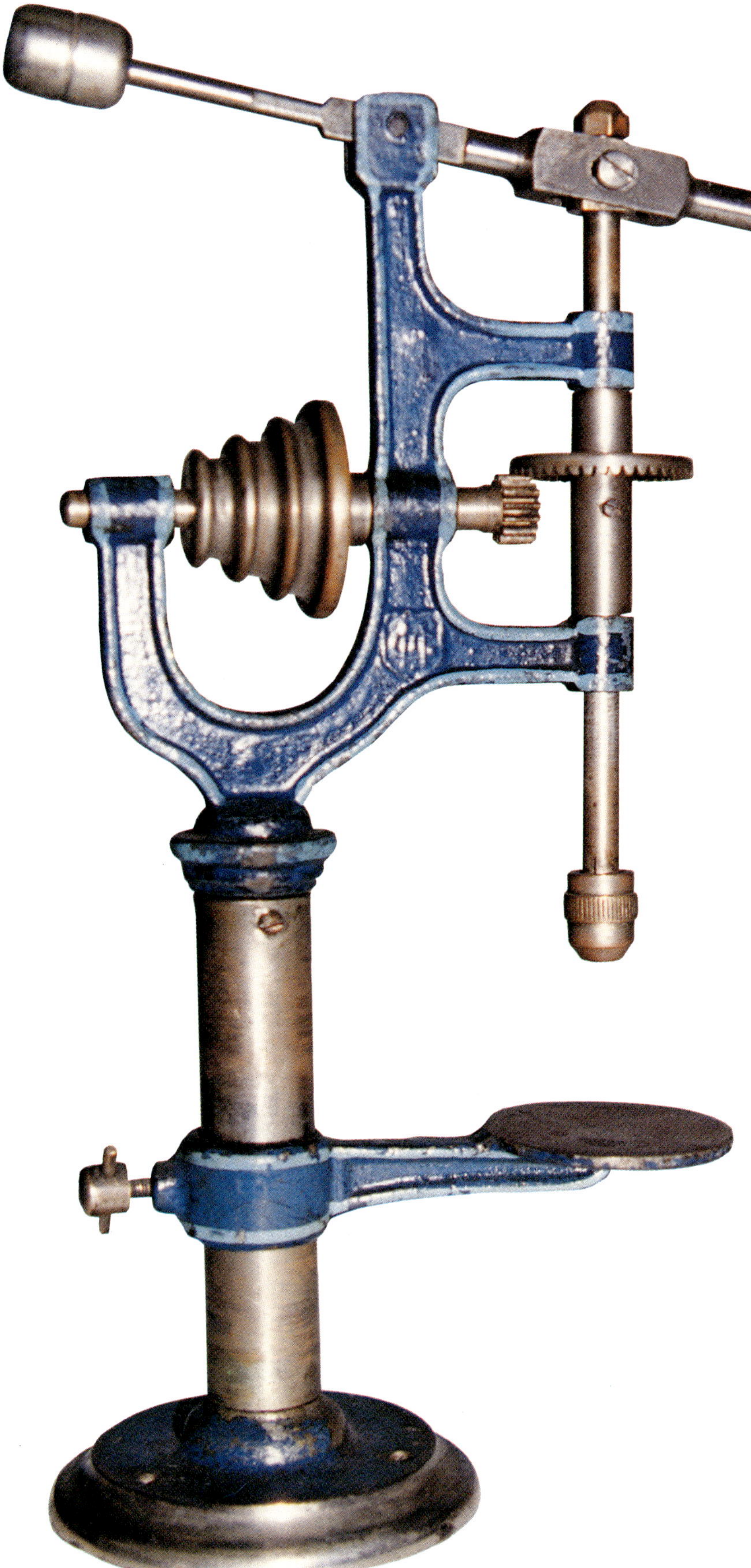

Märklin. Drill press *(Bohrmaschine)*, #4263/2, circa 1909. 18.4 x 7 x 25.4 cm. $400-600. *Courtesy of an anonymous collector.*

Märklin. Band saw *(Band Säge)*, #4255, circa 1904. This example is a later vintage of the earlier toy, probably made around 1930. 5.1 x 8.9 x 13.3 cm. $500-800. *Courtesy of an anonymous collector. Photograph by Françoise R. Hirschberg.*

Märklin. Circular saw *(Sägenschärfmaschine),* #4268, circa 1931. 6.4 x 7.6 x 11.4 cm. $500-800. *Courtesy of an anonymous collector. Photograph by Françoise R. Hirschberg.*

Märklin. Drop hammer *(Fallwerk),* #4267, circa 1909. A later vintage of the earlier toy, made probably around 1930. 8.9 x 9.5 x 19 cm. $1500-2000. *Courtesy of an anonymous collector. Photograph by Françoise R. Hirschberg.*

Märklin. Punch press *(Exzenterpresse),* #4265/2, circa 1904. A later vintage of the earlier toy, made probably around 1930. 8.9 x 7.3 x 18.4 cm. $800-1200. *Courtesy of an anonymous collector. Photograph by Françoise R. Hirschberg.*

Märklin. Machine shop, #4281, circa 1904. Shop has boiler, drill press, grinder, lathe, punch press, and overhead transmission line. All of the components of this toy were also sold as separate pieces. Wonderful. 59.7 x 24.8 x 30.5 cm. $2500-5000. *Courtesy of an anonymous collector. Photograph by Françoise R. Hirschberg.*

Märklin. Forming machine *(Metall-Hobelmaschine)*, #4287, circa 1909. 12 x 8.9 x 11.4 cm. $2000-3000. *Courtesy of an anonymous collector.*

Märklin. Vertical saw *(Decoupiersäge)*, #4309, circa 1909. Has vertical and horizontal motion. 13.3 x 13.3 x 21.6 cm. $1500-2000. *Courtesy of an anonymous collector.*

Märklin. Turnip grinder *(Mahlmühle)*, #4337, circa 1911. 12 x 7.3 x 10 cm. $250-400.

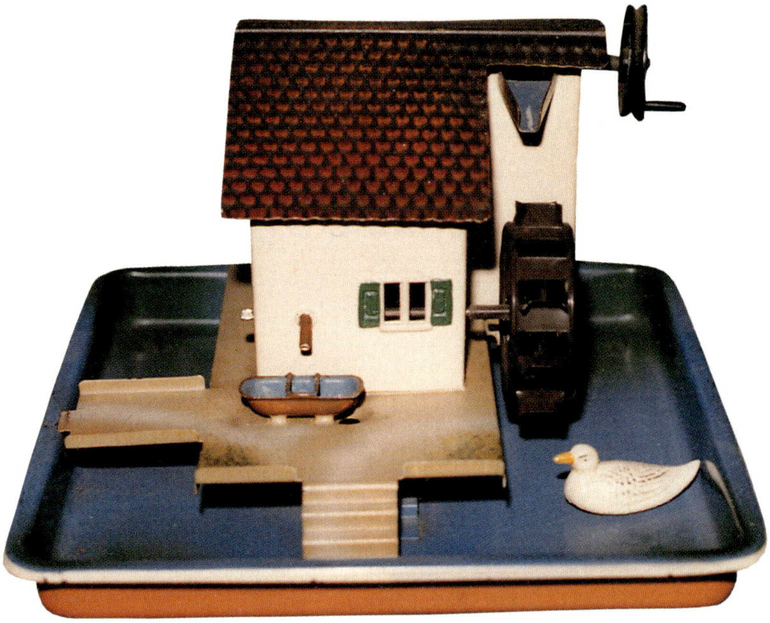

Märklin. Reservoir with mill *(Bassin mit Mühle)*, #4353, circa 1927. 18.5 x 18.5 x 11.5 cm. $400-600. *Photograph by Françoise R. Hirschberg.*

Märklin. Reservoir and house at base of mountain *(Bassin mit Berglandschaft)*, #4352, circa 1927. 20 x 18.7 x 8.5 cm. $400-600.

Märklin. Reservoir *(Teich mit Entenhauschen)*, #4354, circa 1927. When in motion, the mechanism in the house swirls water in the reservoir. 18.7 x 18.7 x 11 cm. $400-600.

Märklin. Reservoir with water wheel and trip hammers *(Bassin mit Felspartie)*, #4355, circa 1927. A shed, which is supposed to be to the right of the water wheel, is missing. 18.7 x 18.7 x 13.4 cm. $200-350.

Märklin. Butter churn *(Buttermaschine)*, #4365, circa 1927. 13.9 x 8.1 x 11 cm. $200-350.

Märklin. Crane, circa 1930. Superb condition. Clutch mechanism to reverse direction. Base is 10.6d cm. Length 22.5 x width 22.8 cm. $1200-1500.

Lot of three dredgers. (At left) Falk, #299/28, circa 1930. 12 x 7.8 x 10 cm. $25-50. (At center) Märklin, circa 1930. 12.3 x 7.1 x 11 cm. $50-75. (At right) Carette, circa 1930. 13.2 x 8 x 13. $25-50.

Märklin. Ferris wheel *(Russische Schaukel),* #8843 and #1123 and #1124, circa 1902. The finest example among toy Ferris wheels. Toy has six boat gondolas with two seats each. At the bottom is a box with a musical movement. 33.2 x 20.3 x 53.2 cm. $50,000 and up. *Courtesy of Jeanne and Bill Bertoia.*

Märklin. Fountain, circa 1930. Walking beam pump, lion's head holding the spout, chain link fence. 19 x 10.5 x 21.5 cm. $500-700.

Märklin. Rear view of fountain showing gargoyle on pump.

Märklin. Fairground carousel *(Karussell mit Musikwerk)*, #8847 and #1127, circa 1902. Truly spectacular. This toy has four sailors on horseback and four scalloped seats, each with two figures. Figures sit around pedestal as well. Box in front is the musical movement. Base is 38.1 cm in diameter. The overall diameter is 45.7 cm, while the height is 60.9 cm. $80,000 and up. *Courtesy of Jeanne and Bill Bertoia.*

Marx

United States

Louis Marx purchased his former employers' company, Strauss, Man & Company of New York City, in 1920. He continued to acquire other companies in the 1920s, and during the 1930s was the largest toy manufacturer in the world. By 1955, Louis Marx & Company had operations around the world. The Linemar line was produced in Japan. The Marx company continued until 1972, when it was sold to Quaker Oats.

A mark used by Marx circa 1960.

Marx. Man at grinding stone, #J-9058, circa 1960. Linemar made only a few steam accessories. 8.5 x 8.5 x 9.5 cm. $25-50.

Müller and Kadeder

Nürnberg

M&K produced a wonderful array of merry-go-rounds or roundabouts. It was M&K which made the boys in the boats (with propellers). In addition, M&K made a number of loop-the-loop toys. M&K operated around 1910.

A mark used by Müller and Kadeder.

Müller and Kadeder. Carousel with four boys in four boats, #135, circa 1909. Two-tiered version. Plays several notes as the toy rotates. Replaced canopy; canoes are missing propellers. This toy and its variations were also made as wind-ups. 24d x 33 cm. $750-1000.

Plank

Nürnberg

Ernst Plank founded his firm in 1866. Very early, he became involved with steam engines and steam toys. The earliest accessories were tools, usually mounted on a wooden base numbered in pencil on the bottom. Later steam toys were carousels and swings. The firm was in business until 1934.

Plank. Walking beam pump *(Pumpwerk)*, #425, circa 1925. The barrel on the left is made of wood. 18.2 x 8 x 15.5 cm. $200-300.

Plank. Walking beam pump with windmill *(Pumpwerk mit Windrad)*, #427, circa 1925. 21 x 8.4 x 29 cm. $400-600.

Plank. Fountain *(Fontäne),* #450/1, circa 1925. 17.5d x 9 cm. $25-50.

Plank. Trip hammer, #416, circa 1902. Mounted on wooden base. 10.9 x 7.1 x 9.5 cm.$50-100.

Plank. Trip hammers, #1010, circa 1914. 10.5 x 7.3 x 11 cm. $50-100.

Plank. Hammer works under roof *(Federhammer mit Dach),* #435, circa 1925. 17.4 x 15 x 16.8 cm. $200-300.

Plank. Fountain with female figure *(Fontäne mit Windkessel),* #450/2, circa 1925. 21d x 18 cm. $200-300.

Plank. Sawmill *(Baumsäge mit beweglichem Baumstamm),* #466, circa 1925. 21 x 13 x 18 cm. $300-500.

Plank. Drill press with control board *(Bohrmaschine)*, #1002B, circa 1914. Made of iron. Control board is paper lithography. 11.5 x 9 x 12 cm. $200-300.

Plank. Sandmill *(Sandmühle)*, #1016/2, circa 1914. 12 x 12 x 25 cm. $150-250.

Plank. Windmill, circa 1914. Note the similarity to the sandmill. 12 x 12 x 26.6 cm. $150-250. *Photograph by Françoise R. Hirschberg.*

Plank. Band saw *(Vertikalsäge),* 1019/1, circa 1902. Mounted on wooden base. 7.6 x 12.7 x 12.7 cm. $300-400. *Courtesy of anonymous collector. Photograph by Françoise R. Hirschberg.*

Plank. Carousel *(Luftschaukel),* #1033/4, circa 1914. There are four seats with a man and woman in each seat. Some replaced figures. 20d x 24 cm. $750-1000.

Plank. Ferris wheel *(Russische Schaukel),* #1024, circa 1902. There are two sailors in each seat. A rare and beautiful piece. 14.8 x 14.8 x 40 cm. $2000-3000.

Plank. Ship swings *(Schiffschaukel),* #1033/5, circa 1914. There are two figures in each of the four seats. Replaced figures and replaced ball ornamentation. 25.7 x 12 x 24.5 cm. $1000-1500.

Plank. Pump station *(Pumpwerk),* #1045/2, circa 1914. Three walking beam pumps: two for the reservoirs and one for the trough. 17 x 13 x 22 cm. $600-1000. *Photograph by Françoise R. Hirschberg.*

Plank. Forced or feed water pump *(Saug- und Druckpumpe),* #1045/6, circa 1914. 7.6 x 8.9 x 14 cm. $300-500. *Courtesy of an anonymous collector. Photograph by Françoise R. Hirschberg.*

Plank. Fountain, #Sortiment Z, circa 1914. Pump has red lion's head. 23.2 x 12.4 x 19 cm. $150-250.

Plank. Marble mill *(Märbelmühle),* circa 1914. Screw mechanism lifts marble to top of tower on the right. Bell rings as marble passes through and later the waterwheel on the bottom turns. 33 x 10 x 28.5 cm. $750-1000.

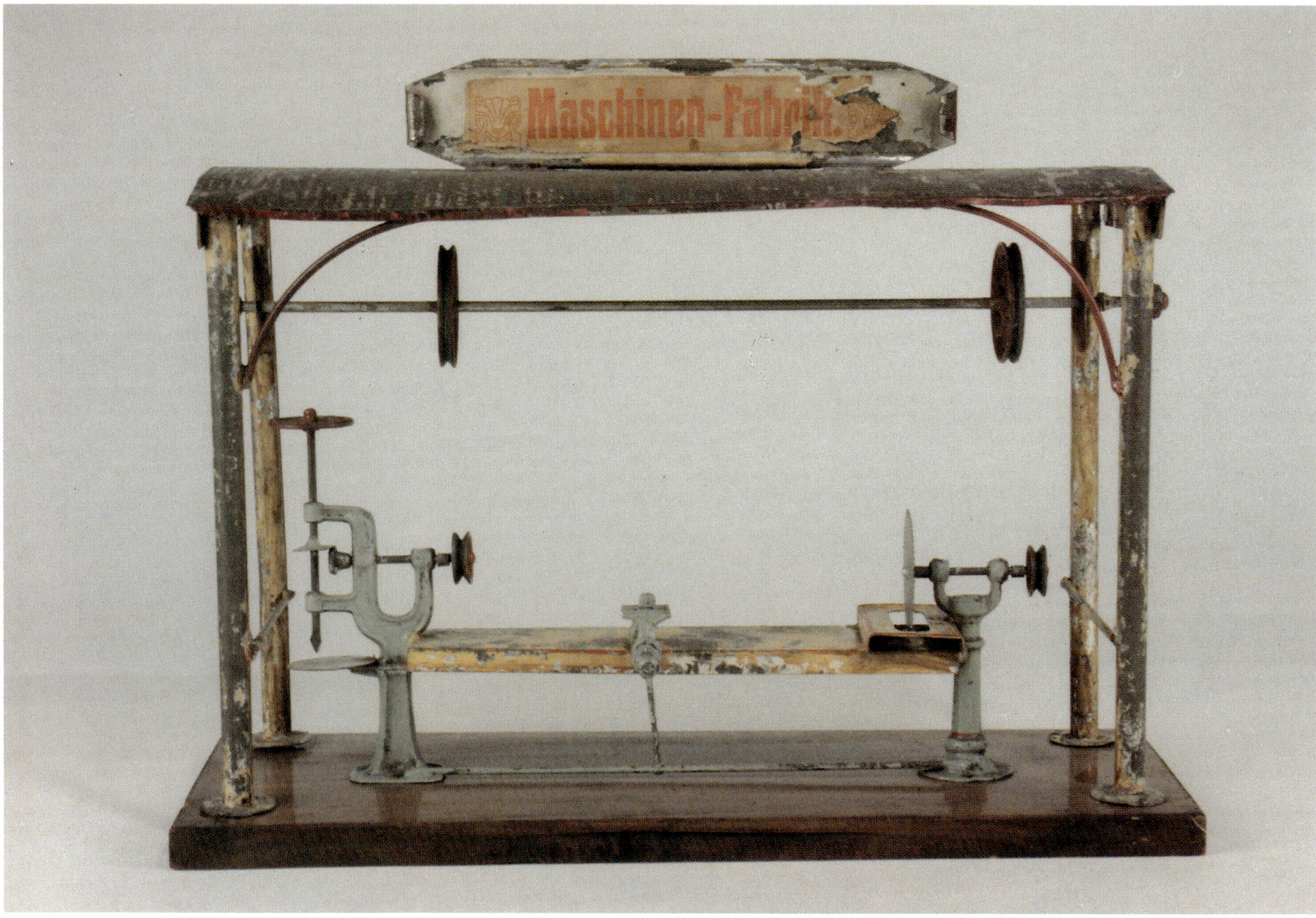

Plank. Machine shop *(Maschinen-Fabrik),* circa 1925. At left is a drill press, in the center a vise, and at right a circular saw, all on a work table. Mounted on wooden base. 30.2 x 11.7 x 23 cm. $250-500.

Plank. Machine shop, circa 1925. At left is a free-standing drill press. Mounted on the work table (from left to right) are a grinder, a vise, and a circular saw. 29 x 15 x 22.5 cm. $500-750.

Rohrseitz

Zirndorf

Karl Rohrseitz began manufacturing toys in Zirndorf in 1881. Early toys were rattles in zinc or brass and nickel-plated. After the World War I, the company began to manufacture tin-plate music boxes, money boxes, and sand pails. His logo was introduced in 1925. In 1930 Rohrseitz began manufacturing lithographed humming tops. After World War II, tops and money boxes were the main items produced.

A mark used by Rohrseitz.

Rohrseitz. Windmill *(Windmühle)*, circa 1948. Marked "Made in Western Germany." Vanes made of plastic. 9d x 15.5 cm. $20-40. *Photograph by Françoise R. Hirschberg.*

Schoenner

Nürnberg

In 1875, Jean Schoenner began manufacturing toys. Among his earliest toys were steam accessories. Schoenner mills are often pinched and look as if someone has crushed them. He produced a number of figural accessories which are clothed and have a Martin quality to them. In 1910 Schoenner closed and part of his production was taken over by Falk.

Schoenner. Horse carousel, #11/2, circa 1875. Two brown and two white horses suspended from metal poles. Metal chains lead to top of toy. Did not come with a canopy. This toy is all original. 21.5d x 29.8 cm. $2000-3000.

Schoenner. Balloon carousel, #11/4, circa 1900. Figures may not be original. 11 x 10 x 19 cm. $2500-3000. *Provenance Perelman Toy Museum.*

Schoenner. Two blacksmiths working at an anvil, #14, circa 1900. 24 x 10 x 15.5 cm. $500-750. *Provenance Gottschalk Collection.*

Schoenner. Blacksmiths shoeing horse, #15, circa 1900. The smith on the left is shoeing the horse, while the one on the right is pumping the bellows and working at the forge. The toy makes a clicking noise, simulating the hammering of the shoe. A toy similar to this was made by Carette. 23 x 9.6 x 13.6 cm. $1500-2000.

Schoenner. Sawyer with bow saw *(Sägemann),* #16, circa 1900. 16 x 6.5 x 16 cm. $300-500.

Schoenner. Guitar player *(Guitarrenspieler),* #23M, circa 1900. Guitar player wags his head. 10.7 x 8.4 x 12 cm. $350-500.

Schoenner. Three workmen, circa 1900. Clothed and full-figured. The man on the left is sharpening a knife, the center man is working at an anvil, and the man on the right is hammering at a forge. 28.6 x 18.1 x 19 cm. $1500-2000.

Schoenner. Blacksmith at anvil, circa 1900. Clothed and full-figured. Missing part of left arm. 9 x 14.6 x 22 cm. $800-1000.

Schoenner. Man with axe at chopping block, circa 1900. Clothed and full-figured. Missing top of head. 8.8 x 14 x 22 cm. $500-750.

Unknown

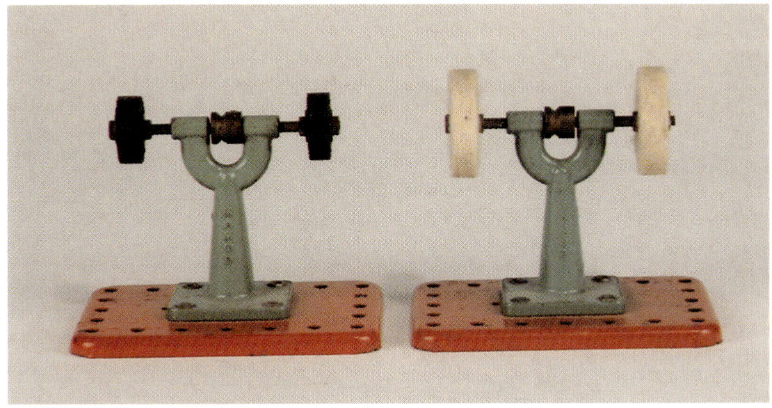

Maker unknown. Two toys, probably Meccano, circa 1950. (At left) Grinding wheels. 9 x 7 x 7 cm. $5-10. (At right) Buffing wheels. 9 x 7 x 8 cm. $5-10.

Maker unknown. Circa 1925. 15.6 x 6.8 x 7.8 cm. $10-20.

Maker unknown. Fan *(Ventilator)*, circa 1930. 11.5 x 11 cm. $25-50.

Maker unknown. Three grinders. (At left) Possibly Carette, circa 1914. 7.4 x 4.2 x 11 cm. $25-50. (At center) Circa 1930. 8 x 6.8 x 9.5 cm. $50-100. (At right) Possibly Zschopau, circa 1905. 11.2 x 7.2 x 10 cm. $25-50.

Maker unknown. Music-maker, circa 1900. Paper lithography face. Musical symbols on body. 8 x 4.5 x 12 cm. $100-300.

Maker unknown. World globe, circa 1900. 7.2d x 16 cm (to top of flag). $100-200.

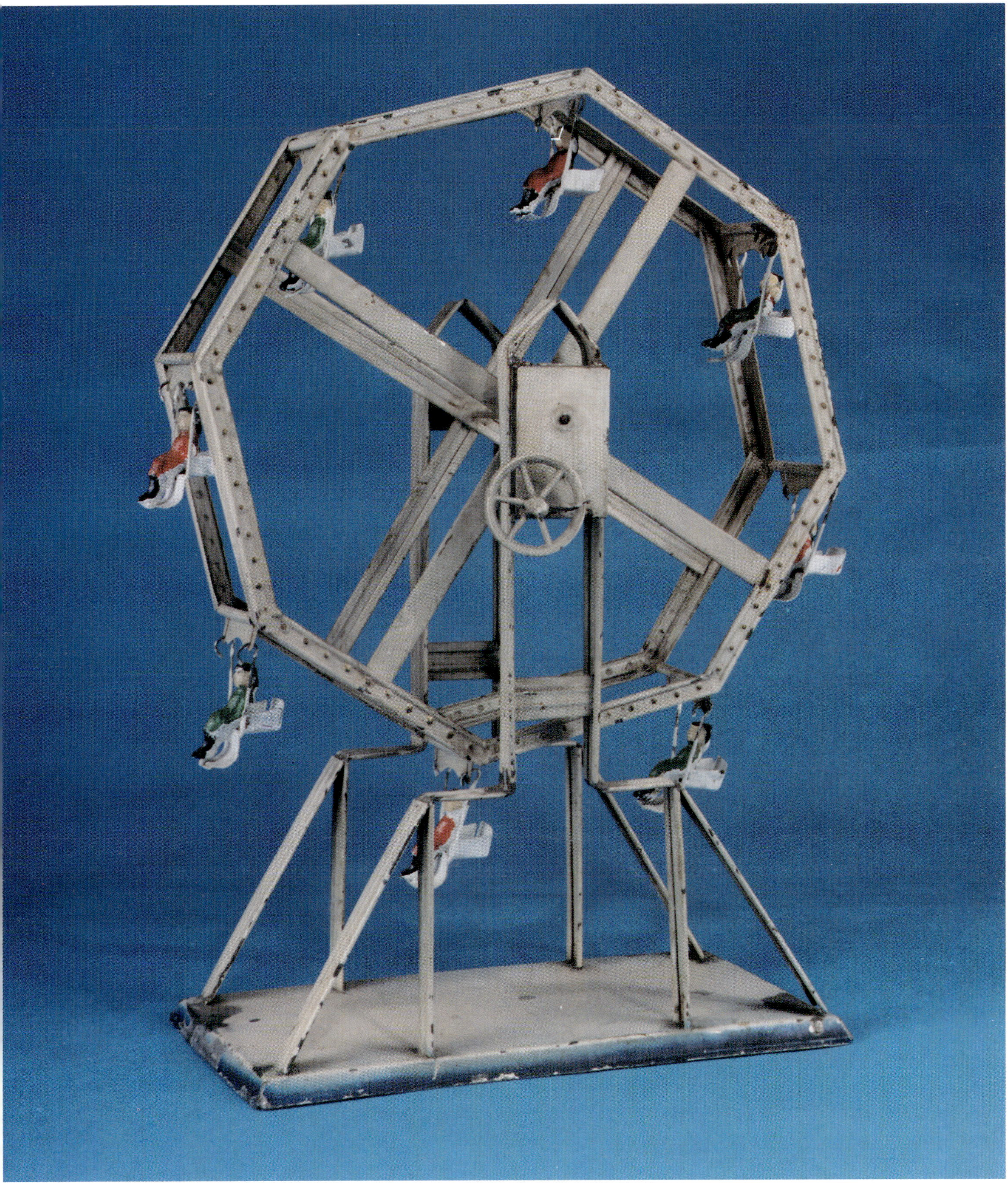

Maker unknown. Possibly Bing. Ferris wheel *(Russisches Karussell),* circa 1930. Replaced seats (note that they are undersized). 25 x 14 x 39.4 cm. $250-500.

Maker unknown. Ferris wheel *(Russisches Karussell)*, circa 1930. Has music box. Sold as Moko 5363/1. 25 x 14.2 x 36.8 cm. $1250-1750.

Maker unknown. Sawyer *(Holtzsäger),* circa 1905. Full-figured. 11.2 x 7.5 x 14 cm. $150-250.

Maker unknown. Chinese sawyer at circular saw, circa 1910. Full-figured. 11.6 x 7 x 17 cm. $300-500.

Maker unknown. Butter-maker, circa 1910. Full-figured. 20.6 x 11.6 x 14 cm. $300-500.

Maker unknown. Possibly Günthermann. Clown drumming, circa 1910. Full-figured. Has cymbals in left hand. 8 x 13.5 x 15 cm. $400-600. *Provenance Gottschalk Collection.*

Maker unknown. Possibly Gunthermann. Clown sitting on drum and beating it, circa 1910. Full-figured with original but tattered clothes. As toy works, musical notes are heard. 10 x 11.5 x 20.8 cm. $350-500.

Maker unknown. Monkey tipping hat, circa 1910. 10.5d x 13 cm. $400-600.

Maker unknown. Circus worker and minstrel, circa 1900. Slate dancers. Man at left holds broom and dust pan. Figures are lead. Background lithography circus scene. 16.8 x 7 x 11 cm. $1000-1500.

Maker unknown. Furnace stoker, circa 1930. 17 x 6.6 x 12.5 cm. $500-700. *Photograph by Françoise R. Hirschberg.*

Maker unknown. Two men in gondolas, circa 1900. 8d x 16.5 cm. $500-750.

Maker unknown. Two dirigibles, circa 1930. Two figures in each gondola. 29 x 12 x 29 cm. $500-750.

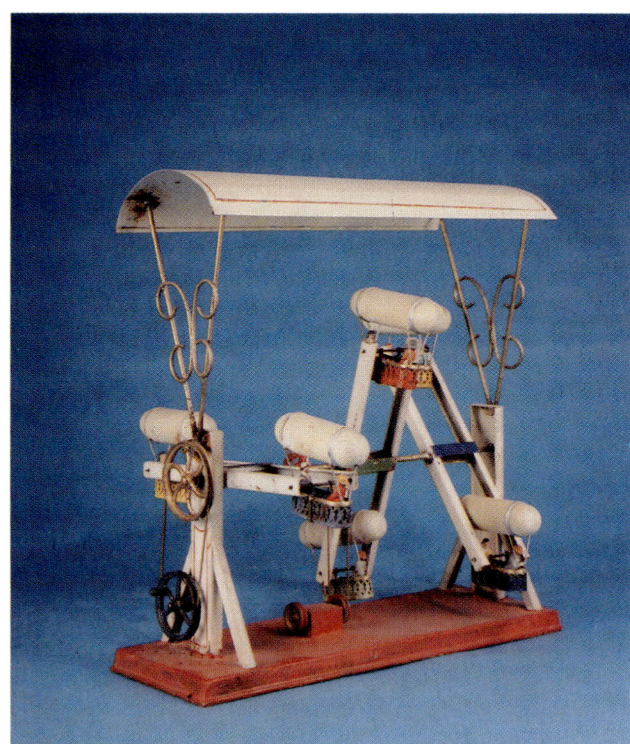

Maker unknown. Three sets of dirigibles, circa 1930. Two figures in each gondola. Same maker as previous two-dirigible toy. Canopy might be an addition. 34.3 x 12.4 x 35.6 cm. $1000-1500.

Maker unknown. Three dirigibles, circa 1910. Men have telescopes. 12d x 26 cm. $1000-1500.

Maker Unknown. Airship carousel, circa 1910. Double canopy. Three balloons, three dirigibles and three airplanes. The dirigibles and airplanes are penny toys made by Distler. It is not known who made the balloons. 31.8d x 44.5 cm. $3500-5000. *Provenance Atlanta Toy Museum.*

Weeden

United States

The Weeden Manufacturing Company operated from 1882 to 1939 in Boston, Massachusetts. Founded by William Weeden, the company produced hundreds of steam engines. They produced only a few metal accessories: a steam plant, a windmill, a saw, and some transmissions. It is hard to miss a Weeden metal toy since they are brightly painted in orange and green. More importantly, Weeden produced six paper lithography toys which are scenic and highly desirable.

Weeden. Circular saw, circa 1930. 9.8 x x 7.5 cm. $15-25.

A mark used by Weeden.

Weeden transmission, circa 1930. 11.5 x 5.5 x 7.5 cm. $10-20.

Weeden. Blacksmiths at forge and anvil, circa 1930. Paper lithography on wooden base. 21 x 2 x 25 cm. $400-600.

Weeden. Windmill, circa 1930. Characteristic Weeden steam toy colors: orange and green. 15 x 15 x 20 cm. $50-100.

Weeden. The Irish Jig, circa 1892. Paper lithography on wooden base. The man twirls his shillelagh while the fiddler plays. 14.5 x 5.2 x 17 cm. $600-800.

Wilesco

Lüdenscheid

A mark used by Wilesco.

The company name was formed by using letters from the founder's name, Wilhelm Schröder. Wilesco began in 1912 and is still in business. Their steam toys were made largely for export and they currently make about six toys which are easy to obtain. Perhaps their most notable steam toy is a model of an atomic plant, which was manufactured in the late 1950s and early 1960s.

Wilesco. Color wheel, current. 11d x 12.5 cm. $25-50.

(At left) Maker unknown. Centrimotor, circa 1930. 8.8 x 8.5 x 6.5 cm. $25-50. (At center) Wilesco. Drill press, current. 9.5d x 13 cm. $20-50. (At right) Probably Falk. Two trip hammers, #286/51, circa 1930. Also Arnold #S15 and Moko #5560. 9 x 6.5 x 7 cm. $20-50.

Wilesco. Cement mixer *(Betonmischer)*, #M63, current. Toy rotates about two axes. Shown with box. 12d x 12 cm. $20-40.

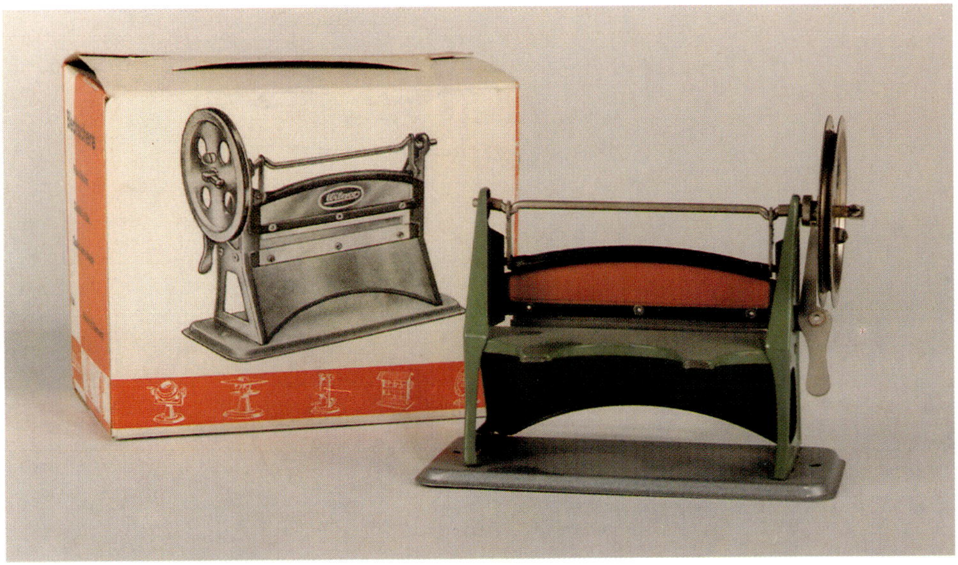

Wilesco. Tin cutter *(Blechschere)*, #M64, current. Shown with box. 13.2 x 8.5 x 12 cm. $20-40.

Wilesco. Commercial washer *(Rollfab)*, #M67, current. Toy rotates and has dinner ware. Shown with box. 13.3 x 5.9 x 8.8 cm. $20-40.

Wilesco. Red and white spiral color wheel *(Lebensrad-Farbspirle)*, #M69-1, current. Shown with box. 10 x 8 x 11.5 cm. $20-40.

Wilesco. Windmill *(Windmühle),* #M68, current. Made primarily of plastic. Shown with box. 11 x 10.6 x 17 cm. $20-40.

Wilesco. Carousel with rocket seats *(Karussell mit Gondeln),* #M70, current. Shown with box. 12.6d x 19 cm. $75-100.

Wilesco. Sawyer with circular saw *(Holzsäger),* #M73, current. Shown with box. 10.7 x 6.4 x 10.5 cm. $20-40.

Wilesco. Scissors sharpener *(Schernschleifer),* #M74, current. Shown with box. 10.7 x 6.4 x 10.5 cm. $20-40.

Wilesco. Man at stamping press *(Stampfer),* #M75, current. Shown with box. 10.7 x 6.4 x 10.8 cm. $20-40.

Wilesco. Sawyer with bow saw *(Schreiner),* #M76, current. Shown with box. 12.4 x 6.4 x 10.5 cm. $20-40.

Wilesco. Man at stamping press *(Stampfer),* circa 1948. 10.7 x 6 x 11 cm. $50-100.

Wilesco. Man at circular saw *(Holzsäger),* circa 1948. 10.7 x 6 x 11 cm. $50-100.

Wilesco. Scissors sharpener *(Schernschleifer),* circa 1948. 10.7 x 6 x 11 cm. $50-100.

Wilesco. Sawyer with bow saw *(Schreiner),* circa 1948. 10.7 x 6 x 11 cm. $50-100.

Wilesco. Nuclear energy plant *(Atomkraftwerk),* circa 1958. Powered by electricity. 30.5 x 25 x 18 cm. $250-500.

Wunderlich

See the Zschopau School

Wunderlich. Clown with hoop in barrel, #143/36/3, circa 1924. As clown rotates he is lowered into the barrel and then rises. Small tear in base and some paint loss. 8.6 x 8.6 x 18 cm. $300-500. *Photograph by Françoise R. Hirschberg.*

Wunderlich. Man at tin cutting machine, #143/54/2, circa 1910. 12.3 x 7.6 x 12.5 cm. $250-400.

Wunderlich. Man at drill press, circa 1910. Man holds hammer in right hand and strikes vise. Drill press attached to work table. 13 x 9 x 13 cm. $250-400.

Wunderlich. Man painting wall, #143/71, circa 1924. 12 x 9 x 12.5 cm. $600-1000.

Wunderlich. Artist, #143/121, circa 1924. Painting is a landscape. 12 x 9 x 12.5 cm. $500-1000.

Wunderlich. Sawyer at circular saw, #143/155, circa 1924. 10.5 x 7.5 x 12 cm. $150-250. *Photograph by Françoise R. Hirschberg.*

Wunderlich. Bootblack, circa 1924. Bootblack is buffing one boot while the other rests on bench. 12 x 9 x 12.5 cm. $600-1000.

Wunderlich. Baker, #143/165, circa 1924. Baker has bread shovel which moves in and out of the oven. 12.2 x 7.8 x 11 cm. $500-750.

Wunderlich. Sculptor, circa 1924. Sculptor holds wooden mallet and is shaping female figure. 12 x 9 x 12.5 cm. $600-1000. *Provenance Perelman Toy Museum.*

Wunderlich. Man with horn plane, circa 1924. 12 x 7.5 x 13 cm. $600-1000.

Wunderlich. Shoemaker, circa 1910. Shoemaker is repairing boot with hammer. 9.3 x 6 x 10 cm. $300-500.

Wunderlich. Pie bakers, circa 1910. The baker on the left is kneading dough while the one on the right is rolling dough. There are two pie tins on the table, the one on the right ready for the oven. 15 x 10.6 x 14 cm. $500-750.

Wunderlich. Miner at wall and crusher with two hammers, circa 1924. The miner that should be operating the crusher is missing. This toy is similar to Becker #83/2. 23.2 x 10.5 x 15.5 cm. $300-500.

Wunderlich. Shoe repair shop, circa 1924. The shoemaker on the left is hammering a boot while the shoemaker on the right is buffing a boot. The mates to the boots being worked on are on the work table. A repaired pair of low-cut boots await their owner. 17.5 x 10.5 x 14.2 cm. $1000-1500.

The Zschopau School

Zschopau

The Zschopau School consists of five manufacturers: Heino Becker, Bernard Hommola, Franz Richter, Franz Primus Wunderlich and Carl Zickner. These manufacturers specialized in the production of steam engines and accessories. Only Becker and Wunderlich will be discussed here. For those geographers reading this, Zschopau is about 18 kilometers southeast of Chemnitz, Germany.

I rely quite heavily on the information in Wolf Kaiser and Carlernst Baecker's research from their book "Blechspielzeug Dampfspielzeug" for the following discussion.

Zschopau toys, which are nearly all figural, were once mostly attributed to Karl Bub. Bub's role with the Zschopau School is not entirely clear. Apparently, the Hommola factory founded near the turn of the century was acquired by one Carl Zickner and taken over by Becker around 1906. Wunderlich, a salesman for Becker, began his own company around this time as well.

The Becker Wunderlich toys were handcrafted and hand-lacquered. Most of what we know about these toys is from a 1910 Becker catalog and later from the 1924-1926 Universal Toy Catalog. After Becker's death, there was a merger between his firm and Wunderlich's. Not much more is known about the Becker Wunderlich connection after the early 1930s.

Becker and Wunderlich apparently had small work forces. However, they did supply toys to Bing, Carette and Plank.

For me, their toys are highly desirable.

Zschopau School. Workman with hammer, circa 1910. Man holds hammer in right hand. Vise on work table. Partially repainted. 14 x 7.6 x 13 cm. $200-300.

Zschopau School. Man with draw knife, circa 1910. Man sits on bench fashioning shingles. Pleasant outdoor scene as background. 12.6 x 8.5 x 12 cm. $350-500.

Zschopau School. Machine shop, circa 1910. A circular saw at left, a drill press at right. 20.2 x 9 x 18.5 cm. $150-250.

Zschopau School. Freight elevator, circa 1910. Rough. 13.5 x 9 x 26 cm. $75-100.

Zschopau School. Dirigibles, circa 1910. A pair of blue and yellow balloons on each side. Riders in gondolas. 13 x 8 x 27.5 cm. $750-1000.

Zschopau School. Automotive carousel, circa 1910. Possibly made by Becker. Three old-fashioned automobiles with drivers rotate while musical sound is heard. 12.8d x 17.4 cm. $1500-2000.

Zschopau School. Sailor in swing, circa 1910. Nicely geared down. 8 x 10.5 x 16 cm. $300-500.

Zschopau School. Girl skipping rope, circa 1910. 10.2 x 7.7 x 16 cm. $500-750.

Zschopau School. Girl jumping through hoop, circa 1910. Girl on right missing. Probably Wunderlich. 13.4 x 6.9 x 16 cm. $100-200.

Zschopau School. Acrobat, circa 1910. Probably Wunderlich. 11.5 x 7.4 x 14 cm. $150-300.

Zschopau School. Acrobatic bear, circa 1910. Probably Wunderlich. 11.5 x 7.4 x 14 cm. $150-300.

Zschopau School. Tandem acrobats, circa 1910. Probably Wunderlich. 11 x 7.2 x 28.2 cm. $300-500.

Zschopau School. Clown with acrobatic dog, circa 1910. Possibly made by Wunderlich. Clown holds switch in right hand while dog stands on swing. 13 x 9 x 18 cm. $1000-1500.

Zschopau School, circa 1910. Black man is a penny toy made by Distler. Girls holding hoop rotate. A bit of whimsy. 14 x 6.5 x 13.5 cm. $350-500.

Zschopau School. Punch and Judy, circa 1910. Punch shakes his head up and down to say yes, while Judy nods no. Hand-painted bodies. 12 x 7.6 x 13 cm. $1000-1500. *Photograph by Françoise R. Hirschberg.*

Zschopau School. Clown, strong man, circa 1910. Clown holds mallet in right hand and strikes plunger to ring the bell on top of the tower. Pear tree in background. Partially repainted. 14 x 5 x 20.2 cm. $750-1000.

Zschopau School. Woman at trough, circa 1910. Woman holds pan. Pump is to her left. A small dove sits on the trough while a goose and lamb move toward a water pan. 11.9 x 22 x 16 cm. $750-1000. *Photograph by Françoise R. Hirschberg.*

Zschopau School. Ferris wheel, circa 1910. All-iron Ferris wheel with four seats. As seats rotate, the bells on the side of the wheel are struck and ring. Repainted. 13.2 x 20.5 x 24 cm. $100-200.

Zschopau School. Carousel, circa 1910. Three swan chariots each with rider. Paint stripped from base. 11.3d x 16 cm. $300-500. *Photograph by Françoise R. Hirschberg.*

Zschopau School. Guarded castle, circa 1910. Four sentries rotate around castle. This toy is similar to Becker #188. 16.8 x 15.8 x 21 cm. $1000-1500. *Photograph by Françoise R. Hirschberg.*

References

Anderson, Curtiss. *Jerry Smith's Collections from the American Past.* Kansas City, MO: Hallmark Cards, 1976.

Antique Toy Collectors of America. *Ives, Blakeslee and Williams Co. Catalog (Reprint).* ATCA: Courtesy of Anthony Annese, 1990.

Antique Toy Collectors of America. *Georg Leonhard Staudt 1888 Catalog (Reprint).* ATCA: Courtesy of Samuel Downey, 1984.

Antique Toy Collectors of America. *Müller and Kadeder Blechspielwaren Fabrik Nürnberg Catalog (Reprint).* ATCA: 1974.

Ayres, William S. *The Warners Collector's Guide to American Toys.* New York: Warner Books, 1981.

Bachmann, M. Ed. *German Toys Der Universal-Spielwaren-Katalog 1924-1926.* Cumberland, MD: Hobby House Press, 1985.

Baecker, Carlernst, Haas, Dieter, Eds. *Die Anderen Nürnberger (Vols 1-5),* with Väterlein, Christian Ed. *(Vols 6-7).* Frankfurt: Hobby Haus, 1973-1987.

Baecker, Carlernst, Haas, Dieter, & Jeanmaire, Claude. *Technical Toys in the Course of Time Marklin Vols 1,2,4.* New York: Hastings House, 1978.

Baecker, Carlernst & Väterlein, Christian. *Vergessenes Blechspielzeug Germany's Forgotten Toymakers.* Frankfurt: Michael Kohl, 1982.

Barrett, Noel. *The Gottschalk Auction Catalog.* New Hope, PA: Barrett, 1990.

Bartholomew, Charles. *Mechanical Toys.* Secaucas, NJ: Chartwell Books, 1979.

Becker, Heino. *Fabrik feiner Metallspielwaren.* Zschopau, GE: Becker, 1910.

Cadbury, Betty. *Playthings Past - A Collector's Guide to Antique Toys.* New York: Praeger, 1976.

Carette, Georges & Co. *Preisliste uber optische, mechanische, elektrische und physikalische Ware, Spielwaren und Lehrmittel.* Nürnberg: Carette, 1902.

Culff, Robert. *The World of Toys.* Middlesex, UK: Hamlyn, 1969.

Darbyshire, Lydia, Ed. *The Collector's Encyclopedia of Toys and Dolls.* Secaucus, NJ: Chartwell Books, 1990.

Doucette, Joseph & Collins, Charles L. *Collecting Antique Toys - A Practical Guide.* New York: Macmillan, 1981.

Ertel, Peter. *Blechspielzeug. Mechanische Spielwaren aus der ersten Hälfe des 20. Jahrhunderts.* Berlin: Transpress, 1991.

Fawdry, Marguerite. *British Tin Toys.* London: New Cavendish Books, 1990.

Ferme, H. A. *Playing Under Pressure.* In "Collecting Toys" Vol. 3, #1, February 1995.

Franzke, Jurgen. *Schuco, Bing & Co.* Nürnberg: W. Tümmels, 1993.

Freeman, Ruth & Larry. *Cavalcade of Toys.* New York: Century House, 1942.

Gardiner, Gordon & Morris, Alistair. *The Price Guide to Metal Toys.* Woodbridge Suffolk, UK: Antique Collector's Club Ltd., 1980.

Gardiner, Gordon & Morris, Alistair. *The Illustrated Encyclopedia of Metal Toys.* New York: Harmony Books, 1984.

Gilbert & Kerbow. *German Toys at Auction.* Spring, TX: Beowulf Fine Arts Ltd., 1994.

Gordon, Bob. *Toy Steam Engines.* Aylesbury, U.K.: Shire, 1996.

Harrer, Kurt. *Lexikon Blech Spielzeug.* Düsseldorf: Alba Buchverlag GmBH, 1982.

Hertz, Louis H. *The Handbook of Old American Toys.* Wethersfield, CT: Mark Haber & Co., 1947.

Hertz, Louis H. *The Toy Collector.* New York: Hawthorne Books, 1976.

Hillier, Mary. *Pageant of Toys.* New York: Taplinger Publishing Co, 1966.

Hillier, Mary. *Automata & Mechanical Toys: An Illustrated History.* London: Jupiter, 1976.

Holland, William G., Spong, Raymond, & Whitton, Blair. *American Live Steam Toys & Their Originators.* ATCA, 1978.

Hopkinson, William P. *Toys and Banks With Their Prices at Auction.* Concord, NH: Rumford Press, 1970.

Huxford, Sharon & Bob, Eds. *Schroeder's Collectible Toys Antique to Modern Price Guide.* Paducah, KY: Collectors Books, 1995.

Jeanmaire, Claude. *Gebruder Bing Die grossen Nürnberger 1902-1904 Technisches Spielzeug aus der 'Guten alten Zeit'.* Frankfurt: Hobby Haus, 1974.

Jeanmaire, Claude. *Gebruder Bing 1912-1915 Spielzeug zur Vorkriegszeit Technisches Spielzeug aus der 'Guten alten Zeit' Händler-Hauptkatalog und Nachträge.* Switzerland: Gut Vorhard, 1977.

Jeanmaire, Claude. *Bing Metall Spielwaren 1927-1932.* Switzerland: Gut Vorhard, 1986.

Kaiser, Wolf. *Dampfspielzeug* (in Puppen & Spielzeug) 6#2 1981 and 7#3 1982. Stuttgart.

Kaiser, Wolf & Baecker, Carlernst. *Blechspielzeug Dampfspielzeug.* 2nd Ed. Augsburg, GE: Battenberg, 1991.

Ketchum, William C., Jr. *Collecting Toys for Fun and Profit.* Tuscon, AZ: HP Books, 1985.

Kimball, Ward. *Toys - Delights from the Past.* Lebanon, PA: Applied Arts, 1976.

King, Constance E. *The Encyclopedia of Toys.* London: Crown, 1978.

King, Constance E. *Metal Toys and Automata.* Secaucus, NJ: Chartwell Books, 1989.

Kurtz, Herman. *Spielwaren und Puppen Stuttgart 1912 (Reprint).* Otto Maier, 1983.

Levitt, Jeffrey S. *The World of Antique Toys.* London: Mint and Boxed, 1990.

Levy, Allen, Ed. *The Great Toys of Georges Carette.* London: New Cavendish Books, 1975.

Longest, David. *Antique & Collectible Toys 1870-1950.* Paducah KY: Collector Books, 1994.

Lowe, Becky & Jay. *Maerklin Bros. & Co. High Class Metal Toys 2nd Supplement Catalog H&J 1906 with Supplementary Catalogs H&J 1905 and High Class Metal Toys 1904 (Reprints).* Bethleham, PA: Lowe, 1986.

McClintock, Inez & Marshall. *Toys in America.* Washington, DC: Public Affairs Press, 1961.

McClinton, Katharine M. *Antiques of American Childhood.* New York: Bramhall House, 1970.

Miller, Robert W. *Price Guide to Toys.* Des Moines, IA: Wallace-Homestead, 1975.

Parry-Crooke, Charlotte, Ed. *Toys Dolls Games Paris 1903-1914.* London: Denys Ingram, 1981.

Parry-Crooke, Charlotte, Ed. *Mr. Gamage's Great Toy Bazaar 1902-1906.* London: Denys Ingram, 1982.

Perelman, Leon J. *Perelman Antique Toy Museum.* Des Moines, IA: Wallace-Homestead, 1972.

Pinsky, Maxine A. *Greenberg's Guide to Marx Toys Volume 1: 1923-1950.* Sykesville, MD: Greenberg Publishing Co., 1988.

Pluntky, Peter. *Mekaniska Leksaker.* Stockholm: Camp &Co., 1974.

Pressland, David. *The Art of the Tin Toy.* London: New Cavendish Books, 1976.

Pressland, David. *The Book of Penny Toys.* London: New Cavendish Books, 1991.

Pressland, David. *Pressland's Great Book of Tin Toys.* Great Britain: Golden Age Editions, 1995.

Remise, Jac. *L'Argus des Jouets Anciens 1850-1918.* Paris: Balland, 1978.

Remise, Jac & Fondin, Jean. *The Golden Age of Toys.* Lausanne, SW: Edita S.A., 1967.

Rooks, Marcus. *Toy Steam Accessories.* Buckinghamshire, U.K.: Shire, 1996.

Saran, Fritz. *Katalog uber Lehrmittel - Hauptliste VI.* Berlin: Halberstaudt-Rathenow, 1916.

Schroeder, Joseph J., Jr., Ed. *The Wonderful World of Toys Games and Dolls 1860-1930.* Northfield, IL: Digest Books, 1971.

Spero, James, Ed. *Collectible Toys and Games of the Twenties and Thirties from Sears, Roebuck and Co. Catalogs.* New York: Dover Publications, 1988.

Spilhaus, Athelstan & Kathleen. *Mechanical Toys - How Old Toys Work.* New York: Crown, 1989.

Tempest, Jack. *Collecting Tin Toys.* London: William Collins Sons & Co. Ltd., 1987.

Walter, Hans-Willi. *Sammler-Katalog Ausgabe 93/94 Marklin Spielzeug 1909-1954 Band 4.* Stuttgart: Stuttgarter Druckerei GmBH, 1994.

Walter, Hans-Willi. *Sammler-Katalog Ausgabe 93/94 Doll Spielzeug/30ER Jahre Band 5.* Stuttgart: Stuttgarter Druckerei GmBH, 1994.

Weltens, Arno. *Mechanical Tin Toys in Colour.* English Ed. Dorset, UK: Blandford Press, 1977.

White, Gwen. *Toys, Dolls, Automata: Marks and Labels.* London: B.T. Batsford Ltd., 1985.

Whitton, Blair. *The Knopf Collectors Guide to American Antique Toys.* New York: Alfred A. Knopf, 1984.

Zabern, Philipp von. *Die Welt aus Blech.* Mainz am Rhein: Zabern, Philipp von, 1981.

Index